The Eclipse of

DARWINISM

WITHDRAWN

'The Eclipse of

DARWINISM,

Anti-Darwinian Evolution Theories

in the Decades around 1900

PETER J. BOWLER

The Johns Hopkins University Press

BALTIMORE *&* LONDON

The Johns Hopkins University Press, Baltimore, Maryland 21218

The Johns Hopkins Press Ltd., London

Library of Congress Cataloging in Publication Data

Bowler, Peter J.
    The eclipse of Darwinism.
    Bibliography: p. 257
    Includes index.
    1. Evolution—History.   I. Title.
QH361.B68 1983   575.01'6   82-21170
ISBN 0-8018-2932-1

For Sheila

CONTENTS

PREFACE

In recent years, there has been a resurgence of opposition to the Darwinian theory of gradual evolution through the natural selection of random variations. The creationists have rejected the whole idea of evolution and, some would say, the whole scientific approach to the past. Even within orthodox science, voices have been raised against the revised Darwinism of the modern synthesis. Alternative mechanisms of evolution are freely discussed, and a few biologists have even suggested that the evolutionary approach may indeed be misguided. This book is not a contribution to these debates, although it may be of some interest to the participants. Any reader expecting to find an analysis of the present status of Darwinism will be disappointed, although in the Conclusion I have tried briefly to draw some brief comparisons with the situation around 1900. In general, the lessons to be learnt from the eclipse of Darwinism do not lie in any direct similarity between the old and the new alternatives. My thesis is that the earlier attacks on Darwinism were inspired largely by the revival of a more traditional philosophy of nature which could never be taken seriously today. The lessons lie, if anywhere, in a deeper understanding of the way scientists behave when they realize that currently accepted theories and values are no longer functioning quite so smoothly as they once did.

My cautious attitude toward the links between the earlier and the modern debates over Darwinism is in part a self-justifying mechanism, since I am not a biologist and have only an outsider's understanding of the modern concepts. In other words, this book was not written by a scientist trying to discover the path leading toward our present level of knowledge. It was written by a historian of science interested in tracing the course of certain themes from the original Darwinian revolution into early modern biology. Its purpose is to show the extent to which biologists around 1900 were still influenced by traditional ways of thought, thereby stressing the extent of the further revolution that was needed to establish the view of evolution most popular today. For this reason I have not attempted to pass judgment on what "really" happened in Kammerer's famous midwife toad experiments, or any of the other supposed demonstrations of non-Darwinian evolution. I am more interested in why Kammerer and his colleagues were so opposed to Darwinism and why, in the end, they were unable to adapt their ideas to the spirit of twentieth-century science. If an earlier generation of Lamarckians did fake or misinterpret their experiments, the real question is not how they did this, but why—and why their opponents chose to see everything in so different a light.

Apart from books with an obvious axe to grind, such as Arthur Koestler's *Case of the Midwife Toad,* modern historians have written remarkably little on the anti-Darwinian theories. The scope of the topic is thus enormous, even when limited to published sources. I cannot, therefore, claim to have done any more than provide a basic outline of the issues involved. My hope is that the present account will serve as a guide and stimulus to other historians of science who will explore these topics in the depth they deserve. If my conclusions are not definitive, they will have served their purpose if they help to generate further research on this fascinating topic.

This book was written while I was teaching at the University of Winnipeg and the Queen's University of Belfast. I am grateful to the interlibrary loan staff at Winnipeg and at the Science Library, QUB, for their help in locating some of the more obscure publications. Many scientists and historians of science, including Ernst Mayr, Stephen Gould, Garland Allen, and Malcolm Kottler, have helped my research with advice and encouragement. Finally, I should express thanks to my wife, who has cheerfully endured no less than four intercontinental moves, all in the interests of preserving my academic career in a time of ever-diminishing opportunities.

ACKNOWLEDGMENTS

Parts of this book are derived from material that has already appeared in the form of articles, and are used with the kind permission of the editors and publishers of the journals concerned: "Darwinism and the argument from Design: Suggestions for a Re-evaluation," from *Journal of the History of Biology* (1977); "Edward Drinker Cope and the Changing Meaning of Evolution Theory," from *Isis* (1977); "Hugo De Vries and Thomas Hunt Morgan: The Mutation Theory and the Spirit of Darwinism," from *Annals of Science* (1978); "Theodor Eimer and Orthogenesis: Evolution by Definitely Directed Variation," from *Journal of the History of Medicine and Allied Sciences* (1979).

The Eclipse of

DARWINISM

1

The Eclipse

and Its Implications

THE CRISIS IN DARWINISM

I N 1894 the British association for the Advancement of Science met in Oxford. Once again Thomas Henry Huxley found himself defending the Darwinian theory of evolution against a spirited attack from an amateur scientist. In 1860, Huxley's deflation of Bishop Samuel Wilberforce had caused an uproar, but had established the right of Darwinism to be considered seriously. Now the attack came from the Marquis of Salisbury, a former prime minister and this year's president of the association.[1] The fact of evolution was no longer in dispute—Salisbury even praised Darwin for his achievement in this respect. Although evolution was unchallenged, the value of the *mechanism* that Darwin had proposed to explain it—natural selection—was another matter. No one, claimed Salisbury, had been able to prove that selection could produce significant change in a species; indeed, the theory seemed impossible either to prove or disprove experimentally. In addition, he pointed to Lord Kelvin's calculation that the age of the earth was not sufficient to allow the immensely slow process of natural selection that Darwinians claimed had produced the results we see today.[2] Huxley defended Darwinism with his usual spirit,[3] but he had nothing solid to set against the apparent mathematical certainty of Kelvin's argument. It would be some years before the discovery of radioactivity would suggest a new source of heat that would prevent the earth's cooling at the rate estimated by Kelvin. In the meantime, the physicist's claim represented a major stumbling block for Darwinism.

The opposition was not confined to the physicists. Many biologists themselves had doubts about Darwinism and did not hesitate to voice

[3]

them or to suggest alternative mechanisms. In 1896 Alfred Russel Wallace delivered a paper to the Linnaean Society arguing that all of the characters that distinguish one closely related species from another have adaptive significance, and hence can have been formed by natural selection.[4] This sparked a controversy in the pages of *Nature,* one of those minor disputes that helps to give an indication of scientific feeling at the time.[5] E. Ray Lankester, himself a Darwinian, admitted that Wallace had exaggerated the extent of utility, but claimed that this did not necessarily undermine the selection theory. The Lamarckian Joseph T. Cunningham defied Wallace to convince him that *any* of the characters normally used to identify species had adaptive value. That Cunningham's position was not unusual was confirmed by W. T. Thistleton-Dyer, who lamented that so many had now turned away from the Darwinian approach. Nevertheless, Thistleton-Dyer disagreed with Lankester's solution to the problem, as did W. F. Weldon, a leading member of the biometrical school which hoped to provide experimental proof of selection. Thus, the remaining Darwinists were unable to maintain a united front against the opposition.

Further evidence of uncertainty can be found in the numerous surveys of Darwinism's status published at that time. Even biologists who were well disposed toward the selection mechanism found it necessary to pause and take stock of the situation. They admitted that there was considerable disagreement over the future course of biology. Three surveys stand out as valuable indications of the state of Darwinism at the turn of the century. In Britain, the last of the three volumes of George John Romanes's *Darwin and after Darwin* appeared posthumously in 1897. In Germany, the first of Ludwig Plate's surveys appeared in 1900. And in America, Vernon L. Kellogg published his invaluable *Darwinism Today* in 1907.[6] Each of these writers was fair enough to admit that some of the objections appeared to be valid and that some of the alternative mechanisms would have to be taken seriously, at least as additions to Darwinism.

Those who were opposed to the selection mechanism had no doubt about the overall trend. Darwinism was on the decline and would soon be eliminated altogether as a major evolutionary theory. This confidence is best illustrated by the title of a German work as translated into English, Eberhart Dennert's *At the Deathbed of Darwinism* (1903).[7] The translation was American, symbolizing the fact that here flourished the most vocal school of neo-Lamarckism, one determined to show that selection was at best only a secondary force in evolution. Yet these reports of the death of Darwinism were exaggerated. Already the new Mendelian genetics was growing in popularity and was showing itself to be implacably hostile to Lamarckism. Eventually it would unite with what was left of Darwinism to give the "modern synthesis" that has guided so much research over the last few decades. At first, however, Mendelism was seen as yet another alternative to Darwinism, and this perspective increased both the level of opposition and the level of confusion. Looking back from the 1940s in his classic survey of the modern synthesis, Julian Huxley expressed the belief that Darwinism had emerged from a crisis. He coined the term "eclipse of Darwinism" to describe the situation at the turn of the century when many biologists had turned their backs on selection.[8]

Although the creators of the modern synthesis were aware that they were rescuing Darwinism from a crisis, there seems widespread ignorance of this fact outside the scientific community. Opponents of modern Darwinism from Arthur Koestler through to the Creation Research Society all like to portray themselves as fighting against a theory that the scientific community has never allowed to be challenged. They see Darwinism as a dogma that has become so deeply entrenched because it symbolizes the materialist ideology to which science has committed itself. Since this ideology is supposed to have grown steadily over the last century or more, they imagine that Darwinism too has become steadily more able to shrug off any criticism as "unscientific." A better understanding of the eclipse of Darwinism will thus raise far more than historical questions. If nothing else, it should alter the way in which some people perceive modern Darwinism by showing that the theory has indeed faced and survived major scientific opposition within this century.

The anti-Darwinians' claim that selection has never faced a scientific challenge clearly helps to create their public image, but some of the blame for this situation must rest with the professional historians of science, who have paid scant attention to the alternative theories of evolution. Although many books have been published on the Darwinian revolution, there is nothing written by a professional historian of science to set against Koestler's *Case of the Midwife Toad.*[9] Most accounts, naturally enough, concentrate on the development and reception of Darwin's original theory. All too often they break off at the time of Darwin's death, leaving the impression that the problems facing the selection theory would soon be overcome. An exception is Loren Eiseley's *Darwin's Century*, which does stress the problems, although it does not mention all the alternative theories.[10] A more complete account by Phillip Fothergill seems to have been largely ignored.[11] Curiously, a good impression of the situation can be obtained from Erik Nordenskiöld's classic survey of the history of biology,[12] which was written in the 1920s by a scientist who himself believed that Darwinism would *not* recover!

At a more specialized level, professional historians of science have analyzed some aspects of the situation in biology at the turn of the century; their work, however, has tended to concentrate on the two theories that would eventually be reconciled by the modern synthesis: Darwinism and Mendelism.[13] This material is certainly valuable, but it reveals a certain whiggishness on the part of those who single out only the aspects of the debate that would eventually bear fruit. Far less attention has been paid to theories such as Lamarckism and orthogenesis, which have played little part in modern thinking on evolution. The specialized literature is thus very one-sided and cannot serve as the basis for a comprehensive evaluation of the eclipse of Darwinism. The present study is intended to redress this imbalance by providing an equally detailed discussion of all the theories that were taken seriously by evolutionists around 1900. It will also try to suggest at least in outline, some of the ways in which a better understanding of this episode will affect our ideas about the nature and implications of biological science.

It is also clear that the historical issues have the ability to affect our thinking today. This is particularly true in the context of the ongoing debate between Darwinism and Lamarckism. In opposition to most of the scientific establishment, Arthur Koestler has ensured that Lamarckism would continue to be seen as the possible foundation for a less materialistic view of life. At present, biologists are once again debating the validity of experimental evidence for the inheritance of acquired characters.[14] Some of the old dogmas in genetics are beginning to break down, and a new, more flexible approach is emerging—one that may include a place for effects that would once have been called Lamarckian. A historical study cannot affect the purely scientific questions involved. If anything of Lamarckism survives, it will be through a new synthesis that transcends both the Darwinian and Lamarckian concepts of earlier decades. Yet because these terms are so emotionally loaded, a historical analysis might help us to clarify our thoughts about their wider implications. Whether Lamarckism has any role to play is for biologists to decide, but both biologists and historians need to think very carefully about the moral and ideological consequences of biological theories. The history of anti-Darwinian theories may well reveal that some of the neat connections drawn by modern commentators vanish when the true complexity of the situation is revealed.

CONCEPTUAL ISSUES

The first step in my study will be to clarify the basic positions open to the biologist who wishes to construct a theory of evolution. This is not quite the same thing as simply outlining the theories themselves, since we shall see that the same theory was often understood in different ways by the scientists who supported it. The actual theories I shall be dealing with fall into five fairly obvious categories, but these conceal a number of more subtle differences that often serve to link apparently different theories, or that divide the same theory into two interpretations. The theories themselves are as follows:

1. *Natural selection,* that is, the preferential survival and reproduction of those individuals born with a slight variation in character conferring some adaptive benefit or some advantage in coping with the demands of the environment. The variations were supposed to be produced by a purely random disturbance of the reproductive system, a process not understood in Darwin's time but subsequently identified with genetic mutation and recombination.

2. *Theistic evolution.* In the early years of the Darwinian debate, some scientists with strong religious convictions suggested that variation was not random, but might be directed toward purposeful ends by the Creator's will. By the end of the century, this approach had been abandoned because the element of supernatural design, if taken literally, would put the cause of evolution outside the scope of scientific investigation.

3. *Lamarckism,* which in the late nineteenth century meant only one aspect of J. B. Lamarck's earlier theory, namely the inheritance of acquired characters. In this case the characters are acquired *during the life of the organism* and are supposed to be passed on to the offspring. In its most popular form, use-inheritance, Lamarckism allows the cumulative addition of bodily modifications created by a new behavior pattern adopted by the organism.

4. *Orthogenesis,* which became the term most commonly used to describe evolution consistently directed along a single path by forces originating within the organisms themselves. These involuntary trends unfold without reference to the demands of the environment and may even lead to extinction.

5. The *mutation theory*, a term popularized by Hugo De Vries to denote the belief that evolution proceeds by the sudden appearance of significantly new forms. The mutations occur at random and are non-adaptive, although many naturalists believed that they would still be able to flourish as new species. The term mutation has subsequently been appropriated by Mendelian genetics to denote the spontaneous modification of a gene. In the modern theory, mutations feed variability into the existing population, but as orginally understood, they created new populations instantaneously separated from the original.

Described thus, the theories look deceptively straightforward. In fact, as soon as one begins to look at them in more detail, the distinctions begin to blur because of conflicting interpretations. It is thus essential to have an alternative way of measuring the conceptual differences involved. The scheme outlined below has its origins in a suggestion by Stephen Gould, who argues that there are certain "eternal metaphors" invoked by our efforts to describe the past history of life.[15] In effect, scientists are confronted by a series of decisions about the fundamental principles they will adopt, each requiring them to opt for one side or the other of a crucial distinction. These alternatives, Gould suggests, are constantly debated backward and forward, without any ultimate resolution being possible. The list of distinctions outlined below is not quite the same as Gould's. I have adopted two of his categories, but my first is put in a somewhat different form, one that seems more appropriate to the structure of the anti-Darwinian debate. Each distinction can be outlined in the form of a question, thus:

1. Is evolution an orderly process in which groups of related forms advance through a regular pattern of development, or is it irregular, consisting of a constantly branching, ever diversifying "tree" of life? Darwin's selection theory was essentially a mechanism of irregular evolution, and it certainly helped to destroy the old idea that life as a whole advances in a regular manner along a unified "chain of being." Yet many later naturalists could not escape the feeling that evolution must, on a smaller scale, be an *orderly* process, and they consistently looked for regular patterns in the evolution of particular groups. Lamarckism might at first seem to be consistent with the idea of irregular, branching evolution, since it can be seen as a simple replacement for natural selection as the mechanism of adaptation. In fact, however, many Lamarckians tried to see the consistent effects of habit as a unifying force that would produce regular evolution. Use-inheritance might thus mimic the kind of goal-directed process required by

those who still believed that evolution must be designed by God. Ortho-genesis was almost by definition equated with linear evolution because it presupposed built-in trends (although the term was occasionally used for small-scale irregular effects if they were nonadaptive). The mutation theory was consistent with the Darwinian principle of irregular branch-ing, stressing to an even greater extent the random nature of variation.

2. Is the process of evolution controlled by the demands of the external environment, or by forces internal to the organisms themselves? Dar-winism is a theory of external control, since the internal process of variation is seen as purely random and hence incapable of directing evolution. Thus, natural selection is strictly an adaptive mechanism—it can only produce those characters that are useful in the struggle for existence. Lamarckism also tended to stress the response of the organ-ism to an external challenge, but again there were differences of inter-pretation. A group of organisms might acquire a new habit in response to changes in the environment, but the habit might then take over and become an internal driving force. Some characters might be acquired through an interaction between the environment and the internal constitution of the organisms, in which case it would be a matter of choice which factor was emphasized. Long-range orthogenesis would require a strong element of internal direction, but even here some naturalists insisted that there must be an environmental stimulus to elicit the effect. Mutations were supposed to come from within the organism, although in no particular direction. Opinions were divided on whether or not the mutated forms were eventually selected by the environment.

3. Is evolution a continuous process, the steady accumulation of minute changes in each generation, or does it occur by the discontinu-ous production of totally new forms? Darwinism was certainly a theory of continuous evolution, and the mutation theory was by definition discontinuous, but Lamarckism and orthogenesis could again be inter-preted in either way. In general, they were both supposed to be theories of continuous evolution, but some naturalists did make efforts to intro-duce an element of discontinuity into them. The great advantage of a theory based on discontinuity was that it allowed one to preserve the "reality" of species as distinct units in classification, avoiding the con-ceptual and practical problems that would arise if a single species were gradually able to subdivide itself into several later ones via a phase of indeterminate separation.

It is obvious even from this brief outline that one must be very careful when analyzing the real structure of the various theories, particularly in the case of Lamarckism and orthogenesis. The post-Darwinian debates were not straightforward arguments about clear-cut alternatives. They ranged across a series of highly complex issues that gave the individual biologist plenty of room for maneuvering even within what was labeled as a single theory. Although the actual details of these early debates on evolution theory might seem to be outdated, Gould may well be right when he claims that some of the more basic issues have not been completely resolved by the emergence of the modern synthesis. Today's biologists still have room for conceptual maneuvers, however much the tools of their trade may have been improved. For this reason it is possible that—since there is continued dissatisfaction with the fundamental value of the modern synthesis—an analysis of the earlier debates might help contemporary observers to clarify their thoughts on where they stand with respect to Gould's eternal metaphors. Whatever the value of such a conceptual analysis, however, neither the historian nor the modern biologist can rest content with this level of interpretation. The emergence of so many rival mechanisms of evolution at the turn of the century raises major problems for the historian of science who wishes to understand the process by which new theories are accepted. Beyond this, we must all face up to the broader implications of the various mechanisms of evolution.

HISTORICAL ISSUES

The historian of science has always faced the problem of trying to provide a balanced interpretation of outdated theories. Constantly tempted to search for concepts that were ultimately incorporated into the modern viewpoint, he or she must fight the urge to dismiss as irrelevant the ideas that are now rejected. In the case of the anti-Darwinian theories, a further problem is created by the existence of a movement that has always insisted that the triumph of selectionism was achieved only through a systematic distortion of the truth. Modern Lamarckians have tended to present themselves as members of an oppressed minority, rather than as the heirs to a solid biological tradition. It is essential for both objective historians and Lamarckians to realize that anti-Darwinian theories were once taken seriously by large numbers of professional biologists. These scientists may have been influenced by moral or religious beliefs, but they were able to provide empirical arguments to support their position. Instead of dismissing their work as mere wishful thinking, historians should try to find out why they were able to achieve scientific respectability at the time—and why they eventually lost that status.

Modern historians of science are trained to avoid the distorting effect of today's values. They appreciate that one must put hindsight aside and try to see a situation as it appeared to those directly involved with it. At the same time, this increased willingness to take outdated theories seriously in a historical context has generated a major problem: it no longer seems quite so obvious that a successful new theory triumphs merely because of its greater factual support. All too often it turns out that the supposedly self-evident advantages of the "correct" theory were not apparent when it was first proposed. The process by which a new theory gains its dominant status is a complex one, involving not only experimental testing but also shifting philosophical concerns and changes in the structure of the scientific community. The eclipse of Darwinism provides a unique opportunity for investigating how an ultimately successful theory may find itself overshadowed for a time by its rivals. Where else is there an example in which a major intellectual revolution was precipitated by a theory containing key elements of the final solution, but in which those elements were temporarily rejected by much of the scientific community before their true significance was recognized?

The best-known attempt to create a general scheme describing the emergence of a new theory is Thomas S. Kuhn's *Structure of Scientific Revolutions*.[16] For Kuhn, a successful theory is one that so dominates the scientific community that, in effect, it defines what is perceived as the only scientific approach to the problem. The theory has then achieved the status of a *paradigm.* A scientific revolution begins when a previously dominant paradigm is undermined by too many irreconcilable facts. This precipitates a state of crisis in which many new alternatives are tried out. The revolution is completed when one of these alternatives at last convinces the whole community of its superiority and becomes the next paradigm. There has already been some debate on whether or not the Darwinian revolution fits this pattern, in which Ernst Mayr has insisted that it represents far too complex an event to be rationalized so easily.[17] Even if (for the sake of argument) one accepts John C. Greene's claim that the emergence of modern evolutionism is the rise of a new paradigm to replace supernatural creation, one must still ask if the pattern of events around 1900 fits the Kuhnian scheme.

The confused situation during the eclipse of Darwinism seems to resemble the crisis state before the consolidation of a new paradigm. Several alternative theories were being promoted, two of which—

Darwinism and Mendelism—eventually combined and became the paradigm of the modern synthesis. This would appear to parallel the situation in that other classic scientific revolution when the emergence of Newtonian physics synthesized the conflicting interpretations of Copernican astronomy put forward by Galileo, Kepler and others. It takes time to complete a revolution, time in which a basic idea such as evolution or heliocentric astronomy can build up a framework of subsidiary hypotheses so it can appear convincing. There is, however, a crucial difference between Copernicus and Darwin. Apart from his heliocentric concept, the rest of Copernicus's astronomy was distinctly medieval in character. He thus emerges as a transitional figure, blurring the distinction between the two paradigms. Yet Darwin's theory of evolution contained many elements that would eventually be built into the modern synthesis. A closer equivalent to Copernicus would be Robert Chambers, whose anonymously published *Vestiges of Creation* sparked the evolution debate fifteen years before the *Origin of Species* with a theory that still incorporated the traditional notion of design. Darwin would thus emerge more as a Kepler or Galileo: someone who came on the scene after the debate had started and contributed an important part of the mechanism that would eventually be built into the paradigm. (I should add that I have not the slightest interest in suggesting that we rename the "Darwinian" revolution.)

The real problem appears upon realizing that Darwinism had apparently enjoyed considerable success before its eclipse around 1900. Indeed, the very term "eclipse" suggests a temporary diminution of previous brightness. One could hardly claim that Darwinism had had paradigm status during the 1870s, since there were many objections to it, yet even its opponents at the end of the century saw themselves as fighting against an entrenched Darwinian orthodoxy. It is this seesawing of popularity that is difficult to fit into any general scheme of scientific growth. Why did the original Darwinian theory gain wide support, despite a number of unresolved problems? Why did the level of opposition increase after the first couple of decades? It may well be that these questions can be answered in familiar terms. Kuhn would not wish to claim that the road to paradigm status is a straight or easy one, and the eclipse of Darwinism suggests that in this case, at least, it was even more crooked than we normally imagine.

Although historians must take into account the kind of evidence used to support or attack the various evolutionary theories, the eclipse of Darwinism cannot be explained solely at this level. All the standard objections to natural selection were formulated during its first decade, yet the theory went on to gain wide popularity. At least one of the alternatives that emerged during the eclipse—Lamarckism—had been under consideration for longer than selection itself. New lines of evidence for Lamarckism and other mechanisms were developed by scientists who were converted to evolutionism by Darwin, but who distrusted his mechanism of natural selection. Thus, although the rival theories were evaluated in terms of their apparent success in dealing with technical issues, there was an imbalance built into the situation by the very nature of Darwinism. Since it was Darwin who converted the scientific world to evolutionism, his theory had a head start in developing its most convincing applications. His opponents took time to develop their alternatives into full-scale rivals. They were also able to exploit the frustrations that built up as it became increasingly obvious that Darwinism could not overcome certain crucial objections. Thus, the eclipse of Darwinism was, in part, an inevitable consequence of the fact that Darwinism was powerful enough to convert most scientists to evolution, while at the same time it contained inherent weaknesses that could not be solved within the existing methods of natural history.

If it was inevitable that alternatives to natural selection would eventually be considered, this still does not explain why many biologists went to the extreme of proclaiming their theories as complete replacements for selection. Opinions gradually polarized to such an extent that mechanisms originally accepted as additions to selection were in the end treated as alternatives to it. The growth of anti-Darwinian feeling needs to be explained in terms of developments taking place within the scientific community and in public opinion. At least three levels of analysis will be required to deal with this question. First, I shall describe the growing specialization of nineteenth-century biology and the creation of new disciplines. Second, I shall look at the social relations between the groups of scientists—how they perceived themselves and were perceived by others. And finally, I shall acknowledge the nonscientific factors that may have influenced the thinking of some biologists, a point that leads directly to the emotional issues that are still with us today.

The structure of the Darwinian debate was inevitably shaped by wider developments within science at the time. Nineteenth-century science as a whole was in a state of growth, with a consequent increase in the level of professionalization and specialization. This paralleled the change from the old tradition of natural history to the new "scientific" biology. For the first time, biologists appeared whose interests were so closely defined by a single discipline (such as paleontology) that they had little understanding of or sympathy with, the problems of other areas. Inevitably, this fragmentation of interests allowed a polarization of theoretical opinions, with each group favoring the theory that seemed best able to solve its own most pressing problems. The need to bridge such divisions was only gradually recognized, and the success of the modern synthesis was in part a consequence of the fact that it was the first evolution theory able to relate itself to all biological disciplines.

The problem of overspecialization was compounded by the appearance of a totally new discipline toward the end of the century: the experimental study of heredity and variation. In part, this may have been a response to the problems of Darwinism that did not seem capable of solution within the old tradition of natural history. More than that, it was a natural consequence of the biologists' desire to establish themselves as modern scientists, following the example of the experimental physiologists. Both Darwinism and Lamarckism had been conceived within the old tradition, and at first neither was looked on with any favor by the enthusiastic practitioners of the new science. The idea of continuous evolution in any form was at first rejected by experimentalists convinced that the discontinuous characters studied by Mendelism were the only form of true genetic variation. The Lamarckians fought a rear guard action in trying to provide their theory with an experimental foundation, but had little convincing success. Thus it was that Darwinism eventually emerged from its eclipse once it could be shown that a more sophisticated interpretation of the new genetics would provide a firmer foundation for selection.

My second level of analysis relates to the behavior of groups within science and requires a study of how theories are presented to the scientific community and to the general public. David Hull has suggested that in order to understand a theory's reception, we must take into account the way in which its supporters present it to the public.[18] In effect, we must pay attention to the "public relations" skills (or lack of them) employed to promote the theory. This is not to say that a good PR team can sell a bad theory, but poor tactics at this level might certainly have an adverse effect on the fortunes of a good one, and might play a crucial role in the case of a theory that is only partially adequate. If the early supporters take up dogmatic stands on technical issues and fall out with one another in public, their chances of success will be slim even if there is a core of truth in their basic idea. In contrast, if they take up a flexible position that will admit a number of partially divergent views, their chances of attracting support will be increased. Even those scientists who have reservations about some parts of the theory might join the group to exploit the more successful areas of application. The early form of Darwinism was certainly flexible in this way, and its later eclipse may well have been encouraged by an increasing dogmatism that alienated all those biologists who retained doubts about selection.

RELIGIOUS AND PHILOSOPHICAL IMPLICATIONS

From the beginning, there were many who found Darwinism unacceptable for other than purely scientific reasons. Religious opposition to the basic idea of evolution diminished fairly rapidly, but the selection mechanism itself was another matter. By reducing evolution to a trial-and-error process, selection undermined the traditional belief that nature is a purposeful system designed by a wise and benevolent Creator. It drove home this point by using life or death struggle, with all its consequent suffering, as the driving force of evolution. To many it now seemed that all living things—including human beings themselves —were reduced to puppets, mere passive systems at the mercy of the blind processes of chance variation and environmental change. Not surprisingly, many scientists shared the general distaste for such a theory. As James R. Moore has shown, the oversimplified image of a confrontation between science and religion has obscured the fact that many scientists found themselves on middle ground in the debate over the implications of evolution.[19] For a while, some tried to retain the concept of supernatural design through a system of theistic evolutionism. The course of evolution was shaped by the production of variations, which was not—as Darwin claimed—purely random, but was directed to purposeful ends by the God who sustained the laws of nature. The implication that certain aspects of evolution could only be explained as the results of God's forethought and skill retained the notion of design, but left the central feature of evolution still outside the scope of scientific investigation.

By the end of the century, theistic, or designed, evolution was no longer under serious consideration by the scientific community. At least one central principle of Darwinism had been universally accepted: that evolution was to be explained solely in naturalistic terms, leaving no room for the supernatural. Natural selection was not, of course, the only conceivable natural mechanism. Among those who still felt the force of the moral or philosophical objections to Darwinism, there was a strong incentive to explore alternatives that might do less violence to their beliefs. The great advantage of Lamarckism was that, although apparently naturalistic, it came very close to satisfying the requirements of a morally acceptable alternative to selection. Lamarckism did not require struggle, since all individuals adapted themselves to new conditions and passed on their acquired characters to their offspring. Instead of living things being reduced to mechanical puppets, they could be seen as active, purposeful entities choosing their response to the environment and thus directing their own evolution. Many American neo-Lamarckians saw use-inheritance quite explicitly as the means chosen by a benevolent God to allow life the power of designing itself. The religious message was not always apparent in the writings of European Lamarckians, but Paul Kammerer—of midwife toad fame—also appealed to his theory as an alternative to the mechanistic interpretation of life. Kammerer and many other humanists were also attracted to the prospect that human beings could design their own future by controlling Lamarckian evolution. In the last few decades, Arthur Koestler has kept alive the spirit of Lamarckism by repeating these moral arguments, despite the almost complete opposition of the scientific community. Lamarckism, then as now, appeals because it allows us to believe that the vitality and creativity that most people feel to be the essential characters of life are the real driving forces of nature. That is why the theory could never be destroyed by merely scientific arguments.

Koestler's interpretation of the Lamarckian principle has substantial historical foundations, although it is by no means clear that Darwinism is necessarily as materialistic as many think it is. Whatever one thinks of Lamarckism as science or as a philosophy of life, however, it is clear that Koestler's account of the midwife toad affair presents a distorted image of early twentieth-century biology. Everything is oversimplified to fit the dichotomy between Lamarckian vitalism and the materialism of the genetical selection theory. Kammerer does not even emerge as the last exponent of a powerful scientific tradition, but is presented instead as a solitary martyr driven to suicide by the dogmatic materialism that had already become entrenched in biology. The fact that most geneticists were at first just as suspicious of Darwinism as they were of Lamarckism is never admitted. William Bateson, the geneticist who led the attack on Kammerer's work, is portrayed as a Darwinist, despite his lifelong opposition to the selection theory. If Mendelian genetics and the Darwinian selection theory so obviously complemented each other as expressions of materialism, it is strange that biologists took several decades to recognize the fact. In truth, the story of evolutionism in the twentieth century is far more complex than Koestler would have us believe.

To many biologists, the exposure of Kammerer's apparent fraud with the midwife toad experiment symbolized the bankrupt state into which Lamarckism had fallen by the 1920s. Yet only a few decades earlier the situation would have seemed to be reversed, with Lamarckism itself advancing on many fronts. I have already suggested that a fuller understanding of the eclipse of Darwinism should alter the perceptions of Darwinism's status in modern science. It is not enough to see the selection theory as the inevitable expression of a materialist ideology that has indoctrinated scientists for a century or more. The Lamarckian alternative was seriously tried out in science, but it showed itself in the end to be less capable of generating fruitful research. This is not to say that Lamarckism is absolutely false, only that it turned out to be more fruitful to lay the foundations of a deeper understanding of reproduction and evolution through the alternative of "hard" heredity. Once the foundation is laid, it may well prove possible to extend our understanding to a more sophisticated level at which at least some Lamarckian effects may be acceptable. Biology must advance in whatever way it can as science, even if the most obvious philosophical implications are not to the liking of some. The real danger for any of us, however, lies in oversimplifying our understanding of the relations between biology and philosophy to the extent that we think a scientific experiment must commit us to a moral position. The eclipse of Darwinism reveals that the connections are not so clear-cut. Lamarckism too has its darker aspects, just as Darwinism can be interpreted in a less than completely materialistic way.

Investigation of Lamarckism as it flourished at the turn of the century shows that the optimistic image depicted by Koestler is only part of the picture. Lamarckism meant the inheritance of acquired characters—but there are many different kinds of acquired characters, some of which would not lend support to Koestler's position even if they were inherited. In fact, we shall see that there is a strong historical link between Lamarckism and orthogenesis, the latter being based on an extremely pessimistic concept of an evolutionary process driven by inbuilt forces that may push a species blindly into extinction. By the time historians have untangled this relationship, they will be forced to admit that biological Lamarckism can be expanded in many different ways, of which the optimistic interpretation is only the most superficially obvious. Furthermore, those who objected to the element of struggle in natural selection could also opt for a theory of pure mutation in which evolution is supposed to be totally unguided because any new form produced by chance will be able to establish itself. Between the extremes of evolution absolutely predetermined by internal forces or left completely to chance, both Lamarckism and Darwinism occupy a comfortable middle ground.

Exactly the same point can be made by looking at the ways in which biological theories were used as models for social policies. Richard Hofstadter's classic book generated a popular belief that late nineteenth-century capitalism was built on a foundation of "social Darwinism."[20] The claim that we should abandon all spiritual values and leave everything to the harsh judgments of nature would seem to be a direct extension of the materialist ideology to which Koestler objects. Darwinism has also been seen as the basis for other social policies to which modern liberals object, including eugenics (direct control of human breeding) and the dominant position assumed by certain racial groups. Historians of social theory are now beginning to realize that some of these interpretations are oversimplified. Robert C. Bannister has shown how little biological Darwinism contributed to the social policies named after it, while eugenics in early twentieth-century America drew its biological support from Mendelism rather than Darwinism.[21] When Mendelism and the selection theory eventually merged to give the modern synthesis, one leading exponent of the new Darwinism, J.B.S. Haldane, used the complexity of the theory to argue *against* the direct applicability of biology to society.[22] Further consideration of the anti-Darwinian theories will only reinforce this growing realization that there is no necessary connection between biological concepts and social beliefs.

The claim that Lamarckism would serve as a vehicle for human progress was certainly advanced during the late nineteenth century. Several historians have commented on this optimistic application of the theory,[23] but the term "social Lamarckism" has never gained the popularity enjoyed by its Darwinian equivalent. It is possible that this comparative lack of attention has stemmed from a general feeling that Lamarckism was never anything more than a fringe movement in science, incapable of providing a convincing social analogy. But more important for our present concerns will be a recognition of the widely varying social implications that have been drawn from the Lamarckian principle. According to Hofstadter, the philospher Herbert Spencer was responsible for transmitting social Darwinism from Britain to America—yet the historian of biology sees Spencer as a neo-Lamarckian! The fact that the most prominent exponent of laissez-faire individualism accepted Lamarckian rather than Darwinian evolution suggests that something is drastically wrong with the categories by which we analyze this kind of relationship. In fact, Spencer's support for laissez faire was a product of the popular philosophy of self-help, which does indeed have overtones more of Lamarckism than of Darwinian selection. Thus, the same biological theory could be used as a justification for two mutually opposed social theories.

The image of Lamarckism as a necessarily more optimistic philosophy of life has also been shattered by John S. Haller's revelation that the American neo-Lamarckians were fervently committed to the biological justification of a hierarchy of human races.[24] Stephen Gould has shown that many similar attempts to brand one section of humanity as inferior to the rest were based on the so-called recapitulation theory, according to which the growth of the embryo repeats the evolutionary history of the race.[25] Gould also shows that this theory was associated far more closely with the Lamarckian than with the Darwinian mechanism of evolution. One could make a strong case for treating nineteenth-century neo-Lamarckism as the last bastion of the hierarchical interpretation of nature that Darwinism in principle set out to destroy. Thus, it is simply wrong to treat Lamarckism as the only biological philosophy acceptable to those who oppose materialism and determinism. The historical record shows that this theory is just as capable of generating those harsher interpretations of humanity and society for which Darwinism itself is frequently blamed. If Lamarckism does regain a place in modern biology, let us hope that its supporters learn this lesson and appreciate that their theory gives no guarantee of philosophical or ideological purity. The one lesson that all can learn from the eclipse of Darwinism is that the theoretical structure of any modern science is far too complex to permit finding one-to-one correspondences between its concepts and those wider issues everyone must confront.

2

The Defense of Darwinism

 B EFORE I begin my survey of the alternative theories, it is necessary to explore the role that Darwinism had come to play
in late nineteenth-century biology. How well established was
the theory and what developments took place within it during
the decades after its first publication? What were its strengths
and weaknesses when measured against the biological knowledge of the time? Which groups of scientists were most likely to retain a
loyalty to the selection mechanism, and which would be most keenly
aware of the objections to it? A substantial range of detailed studies in
the existing literature makes it possible to answer these questions and
to build an outline of the changing fortunes of Darwinism in the late
nineteenth century.

THE ORIGINS OF DARWINISM AND ANTI-DARWINISM

Charles Darwin's *Origin of Species,* which appeared in 1859, was
not the first attempt to popularize a scheme of continuous organic
development. Fifteen years earlier, Robert Chambers's anonymously
published *Vestiges of the Natural History of Creation* had argued for
a law of development that gradually pushed life up the scale of organic
complexity toward man.[1] The book had been roundly condemned by
the scientific community for its lack of technical sophistication, and
indirectly for its religious implications. Nevertheless, there can be no
doubt that it aroused widespread public interest in the possibility that
alternatives might be sought to the direct miraculous creation of
species. The *Origin* refueled that interest, but proposed a totally different mechanism of evolutionary change, one that had even more radical
implications.

Having been convinced of the truth of transmutation by the results

of his voyage around the world on H.M.S. *Beagle,* Darwin had begun the search for some natural mechanism that would explain the process. After a number of false starts, he came to see artificial selection as a possible model upon which to build.[2] The breeder picked out those animals with some natural variation that suited his needs and used them to establish a new race or variety. But how could there be a natural equivalent of artificial selection? The answer came through a chance reading of Malthus's *Essay on Population:* in any species, far too many more individuals are born in each generation than the food supply can support; therefore, there must be a struggle for existence in which only those best fitted to obtain a living will survive and procreate. If the environment changes, this struggle will provide a natural means of selecting those adapted to the new conditions, thereby gradually transforming the species. This was the theory that Darwin explored in private for nearly two decades, refusing to publish because of his fear of public outcry. In 1858 his hand was forced by Alfred Russel Wallace, who sent him a paper outlining a somewhat similar concept of selection.[3] Joint papers by Darwin and Wallace were read to the Linnaean Society in the same year, and Darwin immediately began work on the *Origin.*

The theory of natural selection represented a radically materialistic approach to the problem of the origin of species. Traditionally, the adaptation of each form to its environment had been interpreted as a sign of the Creator's benevolence, as in William Paley's *Natural Theology* (1802). Chambers had followed a different concept of design by postulating a divinely ordered plan of creation that gradually unfolded toward ever higher states of existence.[4] In Darwin's theory, however, neither adaptation nor progression could be seen as indications of supernatural control. Although Paley's utilitarian emphasis on adaptation was once again the central feature, adaptation was now a natural process rather than a fixed state requiring supernatural explanation. Selection simply weeded out those variations that could not survive under changed conditions. The variations themselves occurred at random, in the sense that they were produced by a natural process without reference to the demands placed upon the species. In these circumstances, there could be no preordained pattern of development and no necessary progress toward higher forms.[5] Divergence and specialization, not progress, became the fundamental trends of evolution. The development of life could be represented as a constantly branching tree, not as a linear "chain of being." The demands of adaptation in different geographical areas would tend to divide one species into several, each of which would then specialize for its own way of life. In some

instances adaptation might lead to degeneration, as, for example, in the case of parasites, so natural selection certainly did not necessarily lead toward progress. Any species not able to keep up with the constant pressure of a changing environment would dwindle in numbers and finally become extinct.

For all these reasons, Darwin's theory denied all hope of tracing evidence of supernatural design in nature. This extreme materialism naturally stimulated the theological and some of the popular opposition to his system. In addition, there was the problem of the human being's own position in the new evolutionary world view. As Darwin hinted in the *Origin* and stated plainly in his *Descent of Man* in 1871, the human species would have to be considered as just another animal species. The individual's mental and moral powers were not the special gifts of the Creator, but characters already present among the animals that had been enhanced because of their survival value. Even some of Darwin's closest followers found this difficult to stomach, and the general opposition to human evolution was very strong at first. The problem was compounded in Darwin's system since (unlike Chambers's) it did not even give the human being a unique position as the goal of the evolutionary process. In the tree of life, ours is just one branch among many: There is no progressive trend of which we are the inevitable product. Eventually the fact of human evolution was accepted, but many of the anti-Darwinian theories proposed in the next few decades had as at least part of their objective the establishment of some more meaningful position for humanity in the overall scheme of things.

A detailed account of the initial debate over the *Origin of Species* would be out of place in a study of the later opposition to Darwinism. The main episodes in that debate have been described often enough before.[6] My purpose here is to outline the major arguments for and against the theory and to see how it was able to gain initial acceptance. As we shall see, a number of enthusiastic naturalists exploited the theory's strong points, thereby ensuring that the scientific community as a whole was steadily converted to evolutionism. In an account of anti-Darwinian theories, however, it will perhaps not be inappropriate to begin by describing the arguments that were used against natural selection. To begin with, these arguments were developed by naturalists who retained a sympathy for the tradition of supernatural design. They formulated a number of objections that would continue to serve as a major stumbling block for selection. An anti-Darwinian work such as St. George Jackson Mivart's *On the Genesis of Species* (1871) already listed practically all of the objections that would later be exploited

during the eclipse of Darwinism. Mivart's own alternative of super-naturally designed evolution steadily lost ground during the rest of the century, but the objections remained and helped to shape the naturalistic anti-Darwinian theories that would soon come on the scene.

Darwin and his followers, naturally enough, tried to undermine the strength of these objections. Historians have been divided on the quality of their response. Some have seen Darwin as desperately attempting to shore up a fundamentally rotten structure, while others have taken the more sympathetic view that he made the best possible effort to overcome the limitations of his knowledge. Too close a concentration on the problems of selection will certainly distort one's interpretation of the debate. Darwinism expanded because at first its successes received more publicity than the objections, not because it was able to overcome the criticisms leveled against it. From the beginning a threat hung over the theory, waiting until the initial enthusiasm had died away so that naturalists would be able to think more constructively about alternative mechanisms.

The objections were aimed at various aspects of the Darwinian system. Starting with the most general, we can mention the following:

1. The argument from the discontinuity of the fossil record. This had originally been invoked against Chambers's *Vestiges* as an objection to any theory of continuous development. Paleontologists such as Hugh Miller claimed that the sudden appearance of totally new forms at various points in the fossil record could only be explained by miraculous creation.[7] Although Miller died before the *Origin* appeared, the same objection was frequently raised against Darwin's theory and was still being urged by Sir William Dawson as late as the 1890s. Darwin countered by pointing out that the record was obviously incomplete; the gaps were illusions caused by lack of information. New discoveries tended to confirm this by filling in at least some of the gaps, and by the end of the century few paleontologists doubted the basic fact of evolution. However, the gaps that remained convinced some that the Darwinian theory was still inadequate and encouraged the exploration of theories based on mutations or sudden steps in evolution.

2. Lord Kelvin's argument about the age of the earth. Darwin believed natural selection to be a very slow process, so that evolution would require vast amounts of time to produce the effects we observe. He accepted the teachings of Charles Lyell's uniformitarian school of geology, according to which the earth has maintained itself in a "steady state" for an indefinitely long time. In 1868 Kelvin first pointed out that

uniformitarianism was incompatible with the principles of physics: As a hot body, the earth must cool down, and estimates of the rate of cooling suggested that its total age could only be a few million years.[8] This would in turn eliminate any theory of evolution demanding larger amounts of time, particularly Darwin's. Kelvin himself favored some form of theistic evolution in which divine guidance speeded up the process. Later scientists were reluctant to postulate supernatural control, but mutationism and other naturalistic alternatives to Darwinism sometimes received support because they required less time. Only in the first decades of the present century was it finally shown that radioactivity offered a source of energy that had been left out of Kelvin's calculations and would greatly extend the time available.

3. The problem of the existence of nonadaptive structures. Natural selection could develop only those characters that were of some value in the struggle for existence. In 1865 Carl von Nägeli pointed out that there were many specific characters that appeared to have no adaptive value and hence could not have been formed by natural selection (or any other utilitarian mechanism, for example, Lamarckism).[9] Darwin discussed this claim in later editions of the *Origin*, suggesting that in many cases we do not know enough about a species' lifestyle to be sure that any particular character is useless.[10] Furthermore, natural selection could sometimes be forced to develop useless organs because of what Darwin called "correlation"—when the laws of individual growth linked an adaptive and a useless character together too closely, then in order to improve the one, selection must indirectly develop the other. As the debate over Wallace's paper in 1896 shows, this was still a live issue at the turn of the century. Many naturalists still thought that Darwinism overemphasized the utility of specific characters, and they demanded an evolutionary mechanism capable of producing what they thought was a wide range of nonadaptive characters. Orthogenesis and the mutation theory were both anti-Darwinian mechanisms serving this purpose.

4. The problem of the artificial regularity exhibited by some evolutionary developments. Mivart, for instance, argued that the similarities that can sometimes be observed between totally unrelated forms (for example, the eye of the squid and the eye of the mammal) cannot have been brought about by the chance operations of natural selection.[11] Evolution is too regular to be a purely natural process, he claimed; hence it must represent the unfolding of a divinely preordained plan. This was the system of Chambers's *Vestiges*, brought up to date by the elimination of his oversimplified linear concept of the plan's structure.

In later decades many paleontologists though they could see regular trends in the fossil record that similarly could not be explained by the selection of chance variations. The term "orthogenesis" was introduced to denote that evolution was being directed by an internal, regularizing force.

5. The claim that natural selection is not a creative force. Darwin was unable to explain why the variation that distinguishes one individual from another is produced. He offered it as a demonstrable fact, however, that such variation did occur in many different directions, most of which were of no value to the species. Several critics seized upon this point, claiming that as long as the origin of variation was unexplained, natural selection was not a complete mechanism of evolution. It was the variation that was the really creative force, with selection merely weeding out those of its products that were not up to scratch. It was then frequently implied that variation could not be merely random, but must be shaped by positive forces, which thereby direct the course of evolution. Such forces could either be supernatural, as in theistic evolution, or natural, as in Lamarckism or orthogenesis.

6. The argument from blending heredity. This was a totally negative argument against natural selection first advanced by Fleeming Jenkin in 1868.[12] Since Mendel's work was ignored, most naturalists assumed that the offspring normally tended to be a blend or average of all the parents' characters. Jenkin pointed out that if this is so, a single mutated form born with some advantageous character would have no effect on the evolution of the species since its new feature would be swamped by breeding with the mass of unchanged individuals. As a result, Darwin made sure that he no longer gave a significant role to individual sports, but instead based selection on the range of natural variation always present for any character. A few workers continued to opt for mutations, however, and with the collapse of the old theory of heredity at the turn of the century, mutationism came strongly to the fore. It could be treated either as an addition or as an alternative to Darwinism, according to taste. It may be added that Jenkin had also argued against selection's effectiveness in acting upon the normal range of variation, since this always seemed to reach a limit beyond which the species could not change. In a sense, this was eventually confirmed by Johannsen's experiments in the early years of the present century, but Darwin and his followers ignored what was to be Johannsen's crucial distinction between phenotype and genotype. They continued to believe that in the long run selection would be able to act on

the small amount of (somatic) variation left at the limits of the normal range, and could thus break the variation barrier.

Much of the heat in the *Origin* debate was generated by friction between the materialist implications of natural selection and traditional natural theology. Perhaps some of the less enduring scientific objections were inspired merely by the need to defend the old approach to natural history at all costs. In any case, each of the objections described above was still being used decades after the *Origin of Species* first shocked the world. Each had a degree of plausibility that could still command support long after the initial aversion had died down. Considering both the radical materialism of the complete Darwinian mechanism and this barrage of technical arguments, it might seem surprising that the theory gained any wide degree of popularity at all. Yet the evidence suggests that "Darwinism" did become a popular philosophy of nature—even a popular household word—after 1870. As Erik Nordenskiöld's *History of Biology* admitted, "There is no doubt that the power of Darwinism reached its zenith in the seventies and eighties."[13] Nordenskiöld's book itself concluded with the assumption that Darwinism had permanently lost its influence, and his remark confirms an impression that will frequently be reinforced below: that the opponents of Darwinism were reacting against a movement that they considered to have become well established in the biological community. If the criticism never completely died away, it seems to have experienced a lull for some time before breaking out again at the end of the century.

Several reasons can be offered to explain why Darwinism gained this valuable breathing space. First, it must be emphasized that it *was* the *Origin of Species* that converted both scientific and popular opinion to the basic concept of evolution. Chambers's *Vestiges* got the idea into circulation and may have generated a certain amount of support, but opposition from religious groups and even from scientists remained strong. Nevertheless, Alvar Ellegård's survey of reactions to the *Origin* in the British popular press reveals that by 1872 the basic fact of evolution had generally been accepted.[14] Even if opposition to the mechanism of natural selection remained, Darwinism was now able to trade on the reputation it had established as the key stimulus in a great biological breakthrough. For those who did not inquire too deeply into the technical details, Darwinism and evolution became synonymous. This has certainly tended to confuse the issue for historians, but the confusion at the time may have helped to promote the fortunes of true Darwinism. Even uncritical public acceptance must have confounded the opponents and encouraged the supporters of selectionism.

If the scientists were converted to evolutionism at the same time as the general public, their reactions to Darwinism were more sharply polarized. Here we may look to another factor that helped the selection theory—the eventual triumph of the naturalistic viewpoint. In surveying the general conversion to evolutionism, Ellegård notes that there was not at first the same enthusiasm for the selection theory. The most widely mentioned alternative during this first decade was theistic, or designed, evolution, as a means of compromising between the new idea and the old natural theology. Chamber's vision of divinely preordained evolutionary progress now became acceptable to those with theological concerns as the best way of stemming the slide into complete materialism. Even a number of respected scientists, particularly those of the older generation, adopted this view. In the end, however, theistic evolution proved itself incapable of attracting the younger scientists, and even of maintaining its popular support. Darwinism's greatest triumph was that it soon established a complete break between science and religion. Even during its darkest days at the turn of the century, hardly anyone seriously suggested anything but naturalistic alternatives to Darwinism. Between 1870 and 1890, the concept of design had been banished from science by all except a few older men. This collapse of what had at first seemed the most obvious alternative left the way clear for Darwinism to gain plenty of ground.

The younger scientists welcomed Darwinism precisely because it served as the starting point for an investigation of the natural causes of organic change. By eliminating design it opened the door to a whole new world of scientific biology, which had hitherto been labeled "supernatural." Even T. H. Huxley had refused to take evolution seriously until Darwin showed him that it was susceptible to natural explanation.[15] Huxley had rejected Chambers's theory because it still restricted the scope of natural explanation, but selection now revealed that there was nothing in nature that could not in principle be subjected to investigation. From this point on, no one coming into biology for the first time could be put off with the claim that certain subjects were forever beyond reach for scientists, and the search for natural explanations became the order of the day. Of course, there were other natural mechanisms for evolution available besides selection, particularly Lamarck's inheritance of acquired characteristics; but in 1860, Lamarck's reputation still suffered from the ridicule that had been heaped upon his theory by Cuvier and Lyell earlier in the century. Natural selection was the new possibility that had actually opened the door to naturalism, so it is small wonder that the scientists who passed through that door along with Huxley should first explore the more positive aspects of the idea that had opened it for them. Later on, perhaps, there would be more leisure to worry about the difficulties.

Another useful virtue of early Darwinism was its flexibility and lack of dogmatism. Had Darwin and his followers insisted on selection as the only acceptable evolutionary mechanism, they would have been far more vulnerable to the arguments mentioned above. It has been argued that Darwin did, in fact, retreat from his early commitment to selection, and in *The Descent of of Man* he admitted that he may at first have exaggerated its powers,[16] but the difference is more one of emphasis. Even in the first edition of the *Origin* he had accepted the auxiliary roles of Lamarckism and directed variation,[17] while the last edition still remained essentially an exposition of selectionism. The fact that Darwin himself was flexible enough to make selection the principal, but never the exclusive, mechanism of evolution ensured that his views could direct evolutionary thought without putting a strait jacket on it. At the same time,the antiselectionist arguments could be defused by simply admitting that in some cases additional factors were indeed required.

This flexibility was in general shared by Darwin's immediate followers. T. H. Huxley took selection as a starting point for his thoughts on evolution and cheerfully accepted the possibility of both sudden mutation and directed variation.[18] Michael Bartholomew has recently argued that Huxley was never really at home with the details of the Darwinian mechanism.[19] A. R. Wallace was far more dogmatic on the role of selection in animal evolution, but came to accept that some form of supernatural guidance has been responsible for human evolution.[20] Darwin's chief American supporter, Asa Gray, was deeply concerned with the problem of reconciling selection and design and eventually qualified his support by accepting supernatural control of variation.[21] In Germany, Ernst Haeckel openly proclaimed in the subtitle of one of his principal works his intention to synthesize the ideas of Darwin, Lamarck, and Goethe.[22] Even August Weismann, later the most dogmatic of selectionists, admitted at least some directed variation in his early writings.[23] Thus, even Darwin's strongest supporters refused to commit themselves totally to selectionism, and it would be incorrect to suppose that Darwinism ever achieved the status of a paradigm defining the whole context of evolutionary thought. At the same time, this very willingness to compromise the basic mechanism of selection allowed a loose form of Darwinism to gain a wide measure of support. Selection was just one aspect of evolutionism— perhaps the most important, but certainly not the only mechanism under consideration.

THE EXPANSION OF DARWINISM

Thus it was that a significant group of naturalists set out to explore the world of evolution within a framework roughly defined by Darwin's insight. That they faced constant criticism from certain sources should not blind us to the scope of their achievements. The new evolutionism opened up a number of areas of biological understanding and thereby established its initial, fairly solid position. These areas of research can be divided into two broad categories: those defined closely by the Darwinian mechanism, and those more clearly associated with the general idea of evolution. In the first category fall the studies of adaptation, geographical distribution, and speciation by isolation. These belonged to the field naturalists, who appear to have been from the start the most responsive to the mechanism of selection. Far more general was the reconstruction of the history of life by means of paleontology and the indirect clues of comparative anatomy and embryology. The genealogies and evolutionary trees that resulted from this work often had no direct connection with Darwinism beyond their general conformity with the laws of divergence and specialization. In the end, some schools of paleontology turned away from Darwinism toward neo-Lamarckism and orthogenesis, but the early contributions of Huxley, Haeckel, and others ensured that to begin with this approach was strongly associated with Darwinism.

Apart from the general survey provided by the *Origin*, Darwin himself discussed a number of specialized topics in the light of his theory. As Michael Ghiselin has shown, each of these studies, however narrow, had some relation to the overall concept of evolution.[24] Included here is Darwin's work on the fertilization of orchids by insects and on insectivorous and climbing plants, all illustrating the process by which various kinds of adaptation have come about. The theory, however, was also extended to the broader topics of protective resemblances and mimicry, and this was its greatest triumph in explaining adaptations. It was well known that some species escape their predators by camouflaging themselves to fit the background against which they live. Wallace and the other naturalists who had worked in the tropics provided examples of far more specialized protection, such as the insects that had come to resemble dead leaves or twigs. Even more startling was the new kind of mimicry recognized by (and ultimately named after) Wallace's old traveling companion, Henry Walter Bates. Among the butterflies of the Amazon valley in South America Bates noticed some edible forms mimicking the bright warning colors that had been developed by inedible species to warn off predators.[25] In a paper written in 1862, Bates described this form of mimicry as a clear illustration of natural selection. In fact, protective resemblances as a whole were taken as valuable proof of the selection theory, since this kind of adaptation could not be produced by Lamarckism. Because animals in general have no control over their color, and hence cannot develop camouflage by effort, only the selective weeding out of those that have not been gifted with the useful color would explain the process.

The importance of mimicry for the selection theory is confirmed by the fact that the phenomenon came under heavy fire during the eclipse of Darwinism. Many of the opponents I shall discuss below dismissed mimicry as a product of overactive Darwinian imaginations or, alternatively, claimed that the resemblances were too nearly perfect to be produced by mere selection. The similarities were due to coincidence or to parallel variation-trends affecting different species. Julian Huxley discussed these objections and mentioned the studies done in the present century that have demonstrated the real value of camouflage and its potential as a subject for selection.[26] In many cases, the objections came from paleontologists and from the new breed of experimental biologist that came to the fore at the end of the century. The field naturalists themselves remained convinced of the value of mimicry and were the one group that refused to abandon Darwinism even in its darkest hour. The writings of Edward Bagnall Poulton, professor of zoology at Oxford, provide a good illustration of this continuing faith in Darwinism held by at least one type of biologist.[27] The laboratory worker, Poulton insisted, has no real conception of the pressures under which every species lives in the wild, and of course would dismiss mimicry. The field naturalist alone knows the real value of such adaptations and refuses to abandon the one theory that can satisfactorily explain them. Thus, one kind of naturalist kept the faith with Darwinism until eventually the experimentalists sorted out their initial confusions over Mendelism and laid a new foundation for selection.

Darwin's theory also illuminated the study of the geographical distribution of species. As he showed in a chapter of the *Origin,* the existing distribution could not be explained by assuming that similar species had been created wherever there was a similar habitat. It *could* be explained, however, if successful forms were able to migrate from their point or origin, adapting to whatever conditions they encountered, until finally halted by impassable barriers such as mountain ranges or oceans. Wallace confirmed this in the islands of Southeast Asia by defining the line that now bears his name, the line separating the areas dominated by Asian and Australian forms.[28] In the Malay Archipelago (modern Indonesia) the line ran through the strait separating the islands of Bali and Lombok, where the sea is deep and would have served as a barrier to the migration of Asian species even during geological periods when the sea level was much lower. The botanists also found Darwin's theory valuable in this context, and one of the earliest writers to support selection openly was Joseph Dalton Hooker with his essay on the flora of Tasmania.[29] Asa Gray applied the theory in the same way to explain the distribution of the American flora.[30]

Associated with the study of geographical distribution, however, was a problem that the Darwinians took some time to solve: the mechanism of speciation, or divergence as Darwin called it. It was agreed that evolution would seldom be a linear process; usually a multiplication of one form into many is involved. The groups of distinctive individuals known as "varieties" were supposed to be incipient species, forms that had begun to diverge but that had not yet reached the point of reproductive sterility with one another. In one sense this claim gave the selection theory a positive advantage, since it implied that there was no hard and fast distinction between varieties and true species. The impossibility of consistently establishing such a distinction had become a scandal among the taxonomists, but now the apparent problem could be explained as an inevitable consequence of Darwinism. Naturalists began to pay more attention to varieties and to study the different degrees of distinction between them. Wallace, for instance, demonstrated in 1864 that the complexity of the situation among the Malayan butterflies was fully in accordance with the selection theory.[31] But how were the varieties formed? Darwin's work on the Galapagos group had shown him that islands represent naturally isolated locations where the production of distinctive varieties will be encouraged through the prevention of intercrossing. Nevertheless, in many other cases varieties are to be seen coexisting in the same geographical area, where no isolation seems to be involved. As Ernst Mayr has shown, Darwin eventually came to neglect isolation as a means of producing divergence, believing instead that ecological specialization in the same area is a more important mechanism.[32] This change of opinion paved the way for one of the first major disputes to arise within the Darwinian system.

Darwin and Wallace both came to accept sympatric speciation (divergence without geographical isolation), but disagreed over the means by which isolating mechanisms arose to separate the varieties into noninterbreeding groups. Wallace maintained that selection itself could produce sterility, but Darwin was doubtful, and in the end most naturalists accepted the need for some auxiliary mechanism. Romanes claimed that Darwin's theory explained the origin of adaptations (by linear evolution), but not the origin of species (by divergence).[33] He proposed the mechanism of "physiological selection," by which variants infertile with the parent form may be spontaneously produced. Alternatively, H. M. Vernon's "reproductive divergence" supposed that the natural tendency of the most similar individuals to breed together would lead to divergence.[34] The one factor that was not at first considered as the source of isolating mechanisms was an initial period of geographical isolation. Such a view had been proposed by Moritz Wagner in 1868, but was generally rejected.[35] Darwin himself opposed Wagner's views, while August Weismann wrote an essay specifically repudiating isolation as a factor in evolution.[36] Unfortunately, isolation was viewed as an *alternative* to selection, not as a possible addition to it, partly because Wagner himself had insisted that evolution could never take place without isolation under changed conditions. An additional confusion was the failure of the Darwinians fully to appreciate the implications of their theory for what Mayr has called "population thinking." Species and varieties were still defined in terms of morphological differences, no effort being made to determine whether or not such differences actually defined separate breeding populations. Indeed, the difficulty of drawing a sharp line between varieties and species was often taken as showing that species were not real entities, but were merely convenient and somewhat arbitrary groupings of similar individuals.

Mayr suggests that the failure of the Darwinians to adopt a populational definition of species and the dispute over the role of isolation left them open to the broader attacks of the later nineteenth century. Romanes might propose his physiological selection as an addition to Darwinism, but to many it seemed a small step from this to mutation or orthogenesis as an alternative mechanism not only for sympatric speciation but also for evolution itself. Curiously, though, it was while these alternatives were being urged most vigorously that the field naturalists at last began to develop a more sophisticated view of speciation. Perhaps, indeed, it was the pressure of the alternatives that forced them to recognize the vulnerability of sympatric speciation and the advantages of isolation. In the 1880s the work of John T. Gulick on the Hawaiian land snails established a very real correlation between varieties and distinct geographical locations (isolated mountain valleys).[37] Mayr has described Karl Jordan's development of a modern populational concept of species at the same time, again coupled with a recognition of the role of isolation in speciation.[38] Jordan was particularly successful in exposing the weaknesses of the various attempts to explain sympatric speciation. Jordan was a Darwinist of the old, undogmatic school, and was still prepared to admit the possibility of some Lamarckian effect in addition to selection. While refusing to be sidetracked by the rigid polarization of neo-Darwinism and neo-Lamarckism, he nevertheless developed the full logic of Darwin's original insight. Although sympatric speciation was not totally abandoned, by 1905 David Starr Jordan could argue that the majority of field naturalists accepted the need for isolation, while at the same time he lamented the lack of interest shown by biologists of other backgrounds.[39] Once again we see the lack of contact between the field and the experimental biologists, with the former making an important consolidation of the Darwinian position just when the latter were rejecting selectionism altogether. Although it was the experimental approach that led to the rediscovery of Mendelism and the emergence of the mutation concept, these were at first seen as alternatives to Darwinism. Even the early phases of the genetical theory of selection were developed without a recognition of isolation, and it was not until Ernst Mayr built upon the work of Bernhard Rensch to produce his *Systematics and the Origin of Species* in 1942 that the two traditions finally reintegrated themselves.

On a wider scale, the evolutionists also felt obliged to attempt a reconstruction of the history of life as they supposed it must have occurred, using fossil evidence where possible and, when that failed, the indirect clues of comparative anatomy and embryology. In some respects they were remarkably successful, but the seeds of anti-Darwinian sentiment were nourished by a number of problems that remained unsolved. Wherever a reasonably detailed series of fossils could be found for a particular group, its life history could be established with some degree of confidence. By the end of the century no paleontologist would have denied that enough of these series had been found to establish the occurrence of continuous evolution beyond all doubt. Whether the evidence suggested *Darwinian* evolution, however, became increasingly subject to question. The more fundamental evolutionary links between the classes and phyla were but slowly illuminated by fossil discoveries, and enlightened speculation substituted for the lack of evidence. Haeckel and the more enthusiastic of the early Darwinians constructed genealogical trees for the animal and vegetable kingdoms, postulating a series of crucial intermediates for which they had no hard evidence. Inevitably, disputes arose over the interpretation even of fundamentals, and since the evidence precluded a final decision being made, the whole process eventually came to be seen as a somewhat disreputable speculation. Darwinism itself thereby fell into disfavor, even though Haeckel, the boldest speculator of them all, was a Darwinian only by the loosest definition. Yet in one sense these efforts were indeed Darwinian: they were firmly based on Darwin's picture of an evolutionary process governed by divergence rather than linear progress, with adaptation to new conditions being at least in principle the driving force. The opposition of later paleontologists to Darwinism involved more than a mere dislike of the selection mechanism itself—via orthogenesis the anti-Darwinian scientists were making a deliberate attempt to reintroduce a linear concept of development, at least in limited areas.

When Darwin wrote the *Origin of Species* he had to include a chapter on "the imperfection of the geological record," defending himself against the charge that the lack of intermediates between known fossil forms showed there had been no continuous evolution. In effect, he was predicting that further discoveries would tend to substantiate his theory by filling in the gaps with forms that fitted the pattern of divergence and specialization. Up to a point Darwin was lucky: The later nineteenth century did see an enormous expansion of paleontological knowledge that revealed a number of relatively continuous series. Martin Rudwick notes the case of Albert Gaudry who, although he objected to the Darwinian mechanism, showed that a series of Miocene fossils discovered in Greece during the 1850s supported the overall evolutionary picture.[40] The fossils were intermediate between known Eocene and later mammals, and in particular Gaudry suggested *Hipparion* as the ancestor of the modern horse. The question of the evolution of the horse seemed a promising line of attack, and it was taken up by Darwinians such as Huxley. At first Huxley adopted Gaudry's view of a European origin for the horse, but eventually he was convinced that a series of American fossils discovered by Othniel C. Marsh (another strong Darwinian) offered a far more convincing picture.[41] Marsh also described a "law of brain growth," which he believed governed the evolution of the Tertiary mammals, in an attempt to substantiate a progressionist interpretation of Darwinism.[42] Other continuous sequences were soon uncovered among both the vertebrates and invertebrates, enough to convince all paleontologists of the truth of evolution. But were the sequences explicable by the Darwinian mechanism, as Huxley and Marsh claimed? Rudwick suggests that most paleontologists had already rejected Darwinism by 1870,[43] but this is even before Marsh's horse discoveries and is probably something of an exaggeration. Initially paleontology was seen as providing valuable support for the loosely defined form of Darwinism, but as evolutionary thought began to polarize toward the end of the century, more deliberate emphasis was placed on anti-Darwinian interpretations. It was increasingly argued that the continuous trends were too regular or linear to be explained by the selection of random variations, and that this suggested instead either Lamarckism or orthogenesis. The motivations that led many paleontologists to stress a linear arrangement of their specimens are an important component of the story of the anti-Darwinian movement.

Even though the fossil sequences appeared too continuous to be explained by the Darwinian mechanism, the paleontologists were relatively unsuccessful in demonstrating the links between major groups. The only significant breakthrough here during the nineteenth century was the establishment of some fossil evidence for the evolution of the birds from the reptiles. Huxley was prominently associated with the identification of *Archaeopteryx* as a bird with strong reptilian features, as well as with the recognition of birdlike characters in some dinosaurs.[44] Marsh described the Odontornithes, or toothed birds, of North America.[45] These fossils, however, only provided evidence of some link between the classes; they did not elucidate the details of the evolutionary process itself. The other vertebrate classes remained unlinked by any fossil evidence, requiring ever increasing amounts of faith on the part of the evolutionist. E. S. Russell notes that the contrast between the continuity of the known sequences and the gaps between the major groups led paleontologists such as Karl von Zittel and Charles Depéret to question whether existing theories would ever be able to cope with the major steps of evolution.[46] In fact, there is little chance that breakthroughs left detailed fossil evidence, but the continued lack of evidence obviously weighed heavily on the minds of some workers and helped to generate more anti-Darwinian feeling.

In the absence of appropriate fossils, many of the earlier evolutionists had been quite willing on the basis of indirect evidence to go ahead with a reconstruction of the history of life. Haeckel's *Generelle Morphologie* showed that Darwinism could absorb existing morphology by equating the typical form of a group with its common ancestor, thereby immediately generating predictions about phylogeny.[47] His *Natürliche Schöpfungsgeschichte* and *Anthropogenie* extended the process to the construction of a complete evolutionary tree of life through extensive use of the "biogenetic law" connecting evolution and embryology. The use of ontogeny to reconstruct phylogeny poses complex problems that, as Stephen Gould has shown, have confused many historians.[48] Strictly speaking, Darwin's theory did not imply that the embryological development of a modern individual recapitulates its evolutionary history. In his *Für Darwin* (1864), Fritz Müller showed that there are two possible modes of variation corresponding to two different relationships between ontogeny and phylogeny.[49] If variation changes the embryological development of the organism so that at some point it strikes off along a new path, then embryos will never recapitulate the adult forms of their ancestors. All one can assume is that community of early embryological development indicates community of descent—and this is in fact the only assumption that Darwin himself made. In contrast, if variation occurs by adding on a stage to the existing development, then otogeny will indeed recapitulate the evolutionary history of the race. Such an interpretation of variation is more consistent with Lamarckism than with Darwinism, since in this case the variation must be produced by the adult animal and then compressed back into embryology to be inherited. It is thus not surprising that Haeckel, the leading exponent of recapitulation, was attracted to a Lamarckian explanation of variation. Haeckel also thought, however, that the variations competed among themselves and were selected, so by the loose standards of the 1870s he was counted as a Darwinian. The fact that recapitulation did become the standard technique for the reconstruction of evolution is probably responsible for the failure of even such perceptive historians as Arthur O. Lovejoy to recognize its somewhat artificial connection with Darwinism.[50]

It became fashionable to sneer at Haeckel's work during the eclipse of Darwinism, and the fashion is still with us today. Certainly, his speculations were bold and his Darwinism none too strict—yet his enthusiasm seems a natural product of the first triumphant phase of the new evolutionism. If Haeckel's biogenetic law was Lamarckian rather than Darwinian, it proved possible for a number of stricter Darwinians to use it to great effect. Gould gives as an example the work of August Weismann on the evolution of caterpillars and suggests that this kind of detailed work in the end brought more credit to evolutionism than Haeckel's broader speculations.[51] If Haeckel's phylogenies were based on purely morphological relationships, with little effort being made to consider the driving force, adaptation, there were others such as Carl Gegenbaur who could temper the speculations with realism.[52] Postulation of detailed adaptive processes was in any case difficult and merely increased the speculative element of the quest. And if Haeckel's approach was all-embracing, so was that of many others. The hard-headed T. H. Huxley offered bold ideas on the connections between the vertebrate classes, while the question of the ultimate origin of the vertebrates became the center of a veritable hive of biological industry. Only later did the recriminations begin, and it was not the fault of Haeckel and the others that the fossil record persistently refused to confirm or deny their ideas.

For each group Haeckel postulated a single ancestral form whose character was stamped on all of its descendants. If no sign of such an ancestor was revealed by the fossil record, its basic form was deduced from the available morphological and embryological evidence. Thus, in the *Generelle Morphologie* he postulated a "protomammal," which arose from the reptiles and from which the whole class of mammals has descended. This was, in fact, standard Darwinian practice—thus, Marsh had postulated a "protoungulate" living as far back as the Cretaceous period in his attempt to explain the origins of his new order of mammals, the Dinocerata.[53] Nor were these dogmatic assumptions. When Huxley argued that the evidence suggested a far more complex situation for the origin of the mammals, in which several different lines had separately acquired the characters of the class, Haeckel acquiesced.[54] There was no actual fossil evidence for this; indeed, Huxley was so strongly aware of the lack of any connecting form between the reptiles and mammals that in the end he postulated a direct evolution of the mammals from the amphibians by means of some as yet totally unknown type.[55] Such ideas could be tested as new fossils came to light: During the 1890s, the work of Harry Govier Seeley on the Karoo reptiles of South Africa indicates that efforts to resolve the issue were still being made.[56] Seeley's papers typify a continuation of evolutionary paleontology at a practical level, independent of the growing scepticism of Zittel and Depéret. But the true story of the "mammal-like reptiles" was not put together until the twentieth century.[57] Considering the general anti-Darwinian sentiment of the later nineteenth century, it is not surprising that some paleontologists abandoned the hope that such gaps would ever be filled.

There were even more fundamental questions raised when the hope of fossil evidence could never be seriously entertained. Haeckel's Gastraea theory, for instance, attempted to account for the origins of multicellular organisms with the embryological analogy.[58] The origin of the vertebrates also received a great deal of attention, especially after Alexander Kowalevsky's work on the development of *Amphioxus* suggested a connection with the ascidian tadpole.[59] Darwin and Haeckel both took up this interpretation, but an alternative was proposed that derived the vertebrates from annelids and that dismissed *Amphioxus* as a degenerate form of no evolutionary significance. No fossil evidence could be expected: Even if the original vertebrates had been suitable for fossilization, they must have lived in Precambrian times, and with the exception of the soon discredited *Eozoön canadense,* no Precambrian fossils had been discovered.[60] Thus, there seemed little hope that the conflict between the various interpretations would ever be resolved, a situation in which dissatisfaction would soon breed. The disillusionment of the young William Bateson shows exactly what happened. Bateson began his career as an evolutionary morphologist working on the origin of the chordates, but his 1886 paper on this topic reveals him already chafing at the limitations of his methodology in a manner anticipating the *Materials for the Study of Variation*.[61] In this later work he openly scorned a tradition that could only postulate that *if* such and such a variation were produced it might be preserved and expanded upon. What was needed was direct knowledge of the kinds of variation that were in fact produced by living things, and thus Bateson passed over to the experimental school and his work in the foundation of genetics. His subsequent efforts to downgrade selection as the mechanism of evolution reveal the dangers of the early approach. The refusal of Haeckel and others to suggest adaptive causes that might lie behind the developments they envisioned may at first have seemed like a prudent restraint from an even wider level of speculation. In contrast, Bateson's training as a morphologist in this tradition had left him without a sense of the demands of adaptation. Once he became dissatisfied with morphological speculation, he naturally turned to a study of the laws of variation, not of selection, in the hope of finding a more secure foundation for evolutionary theory.

BIOMETRY AND NEO-DARWINISM

Bateson was not alone in switching from morphology to the experimental study of evolution, but his advocacy of discontinuous evolution precipitated a major debate over the value of Darwinism. The defenders of the selection theory in this case were the biometricians—although it is curious that both sides of the debate traced their origins back to the pioneering studies of inheritance by Francis Galton. Galton had helped to disprove Darwin's own theory of heredity (pangenesis) by performing blood transfusion experiments on rabbits to show that the blood did not carry hereditary material from the body to the reproductive organs. In preparation for his work on eugenics, Galton now began a careful study of human variation and heredity, pioneering the use of statistical tools in biology.[62] He paid less attention to the physical means by which characters are transmitted and was thereby able to make a significant break with the existing view of heredity. Traditionally, inheritance was seen as a "force" producing stability or continuity of form within a species, while variation was a less powerful force with a distorting effect. Galton pointed the way toward the modern view that the two effects are part of the same process: the transmission and recombination of numerous hereditary factors within the population. His "law of ancestral heredity" provided an explanation of the transmission of discontinuous characters (incorrect by Mendelian standards) whereby the offspring drew one-half of its characters from the parents, one-fourth from the grand-parents, and so on. For characters such as height, which have a continuous distribution within the population, Galton postulated a different effect known as regression, by which the offspring of parents differing from the norm of their species will be closer to the mean. The mean, he insisted, was characteristic of a stable form and could not be affected by normally fluctuating variability. Thus, Darwin had been wrong to suppose that the selection of continuous variation would produce permanent results: Only the appearance of a new focus of regression in a mutation, or sport, could effectively change the species.[63]

Thus, Galton helped to promote the belief in discontinuous evolution, and although Bateson seems to have been converted to this view while studying in America under W. K. Brooks, it is significant that Galton backed his later position.[64] At the same time, Galton's techniques were exploited in a different way by the biometrical school of Karl Pearson and W.F.R. Weldon. Bateson had no mathematical abilities, but to the biometricians it was Galton's statistical approach that opened the way to a more detailed understanding of the mechanism of selection.[65] Pearson showed that Galton had actually misinterpreted his own theories: If continuous variations were subjected to selection for a significant period, there would not be a regression all the way to the old norm for the species when the pressure of selection was removed.[66] Meanwhile, the Royal Society of London had set up a committee to study variation with the new techniques, and under its auspices Weldon undertook the first attempt to provide experimental proof of the efficacy of selection. Like Bateson, Weldon had started his career as a morphologist; in fact, the two had at first been close friends,[67] but now they moved off in separate directions under the stimulus of the experimental approach. Weldon set out to support selection by demonstrating the correlation between death rates and continuous variation, first with crabs and then with snails.[68] His efforts were unconvincing and were criticized even by Pearson, but his later work on snails established a real correlation between minute variations of structure and survival rates in nature, providing positive evidence that selection as postulated by Darwin could really modify the form of a species.

In the end, however, the new experimental approach to the study of evolution generated far more opposition than support for Darwinism. There are a number of reasons for this, not all connected with the rise of Mendelism. First of all, Weldon's work showed the effect of selection only on a very small scale; it could not counter the claim that there was a limit beyond which normal variation could not be pushed. Also, both Weldon and Pearson held views on the philosophy of science that were considerably in advance of their time and alienated them from the majority of naturalists.[69] Since adaptation was not a function that could be objectively defined, Weldon simply correlated death rates with some easily measurable character and gave no causal explanation for the selection. When E. Ray Lankester criticized him for this, Weldon replied that he was using a Humean definition of causation as constant conjunction and was not obliged to speculate further.[70] Eventually he did propose an adaptive process to cover his second experiment on crabs, but the more successful work on snails again provided only simple correlations of shell size with death rate, on the assumption that the useful characteristics involved would be far too complex for study.

However sophisticated Weldon's views on causation, his unwillingness or inability to deal with the facts of adaptation did not appeal to the field naturalists who were the only other source of continuing support for Darwinism. Finally, of course, there is the whole issue of the growing emphasis on the discontinuity of variation and the rediscovery of Mendelism. Bateson's advocacy of discontinuous evolution in 1894 led to a break with Weldon and a bitter dispute even before the rediscovery of Mendel's work. Personal feelings, along with certain methodological objections, led the biometricians to oppose the rise of Mendelism and the mutation theory of Hugo De Vries.[71] In so doing they turned their backs on what to many seemed the most promising new development to appear in biology since Darwinism itself. Apart from De Vries's apparent evidence for sudden mutation, Johannsen's work on "pure lines" seemed to confirm the existence of a limit beyond which selection of normal variations could not go. The change of fortunes is symbolized by Bateson's gaining control of the Royal Society committee on variation and shaping its subsequent reports to his own ends. Weldon died suddenly in 1906 and Pearson, who was a mathematician rather than a biologist, became increasingly more interested in eugenics. Biometry lost its battle to defend Darwinism and the new experimental approach to biology became the source of a temporary, but nonetheless vehement, opposition to the selection theory.

The last major source of opposition that I need to mention also arose from a reformulation of Darwinism in response to the problems of heredity and variation: the work of August Weismann. Originally Weismann had made intensive studies of small-scale evolutionary problems such as the development of markings on caterpillars, taking a basically Darwinian stand but allowing the possibility of some non-Darwinian factors.[72] Then he turned to the study of cytology in the hope of uncovering the mechanism of heredity and thereby clearing up the great unsolved problem of Darwinism. Failing eyesight forced him to turn from experiment to theory, and in the course of the 1880s he gradually elaborated his theory of the germ plasm.[73] The notion of a separate substance responsible for transmitting the information of heredity was an important one, especially as Weismann identified his germ plasm with the chromosomes in the cell nucleus. Unfortunately, his complex views on the behavior of this substance turned out to be incompatible with the laws of Mendelian heredity, and even before this had come to seem highly artificial in the eyes of many naturalists. In the absence of any experimental confirmation for the details of Weismann's theory, it was branded by his opponents as an untestable hypothesis verging on pure speculation.

The one thing that Weismann could derive from his theory was a sharper definition of Darwinism. The germ plasm enshrined the principle of "hard" heredity; that is, the complete inability of the body to influence the genetic information passed on to the next generation. In a sense, the body, or soma, was only the "host" for its own germ plasm—it carried and nourished the germinal material and had even been derived from it, but it could never affect that material and hence had no control over its own offspring. Thus, Lamarckism was theoretically impossible, a position that Weismann set out to confirm with his famous experiment of cutting off the tails of mice for several generations. The mice continued to be born with full-sized tails, confirming—at least to Weismann's satisfaction—that even a character systematically acquired by the adults of many generations could never become hereditary. He now purged Darwinism of all its original Lamarckian connections and proclaimed natural selection as the only mechanism of evolution. The germ itself might vary because of internal factors, and selection of such variations would constitute the only means of permanently changing the character of a species. To begin with, Weismann insisted that germinal variations were purely fortuitous. Eventually he had to qualify this by admitting a form of "germinal selection" that might give a small amount of directed (nonrandom) variation from within the germ,[74] but for all practical purposes Weismann continued to insist on the "all-sufficiency of natural selection."[75]

The original, flexible Darwinism had explicitly allowed for a Lamarckian component in addition to natural selection. Weismann's more dogmatic selectionism became known as "neo-Darwinism," and it rapidly polarized the scientific community into two mutually hostile camps. A few naturalists did refuse to be drawn into the germ plasm debate, retaining the old flexibility while they tackled some of the outstanding problems that might still be solved in terms of the traditional Darwinian approach. Karl Jordan is one example I have already mentioned; he continued his work on isolation without committing himself to a particular mechanism of transmutation. Even G. J. Romanes, a leading Darwinist and a careful student of Weismann's position, refused to abandon Lamarckism altogether,[76] but attention increasingly focused on those who adopted Weismann's rigid stance. A. R. Wallace had become known as a dogmatic selectionist (except for human evolution) and was soon identified as a leading British neo-Darwinist. Samuel Butler even suggested that this position be called "Wallaceism."[77] E. Ray Lankester, who had written sympathetically of Haeckel's Lamarckian mechanism of heredity in 1876, now disclaimed all connection with the inheritance of acquired characters.[78] Such an action was typical of the new, less tolerant attitude that was becoming associated with Darwinism in the last decade of the century.

Those who harbored doubts about the total efficacy of selection, or who were explicitly committed to Lamarckism, were now forced to take up a position against Weismann and hence against Darwinism. The germ plasm theory polarized opinions by requiring that one be absolutely for it or against it, and those who were against it now felt it essential to redouble their efforts against selection as well. To discredit Weismann it was necessary to stress more strongly than ever before the evidence for Lamarckism, and opposition to the germ plasm theory became a central feature of the increasingly vocal neo-Lamarckian movement. Herbert Spencer's philosophy of evolutionism had always contained a strong element of Lamarckism, but he had not been seen as an opponent of the original Darwinians. Now he was forced to turn against Weismann and to show the inadequacy of the selection theory in order to retain a role for acquired characters.[79] His arguments were seen as a central platform of the new, more militant neo-Lamarckism, and were buttressed by additional points derived from the work of a host of naturalists in several different areas of specialization. What had begun as an effort to put Darwinism on a firmer footing had once again failed, and had helped to turn anyone with significant doubts against the theory altogether. To a large extent it can be argued that the eclipse of Darwinism was precipitated by movements *within* Darwinism that created a situation in which lukewarm supporters of the theory were turned into active opponents. No longer could Lamarckism and the other alternatives be seen as additions to selection; they would now have to stand up for themselves as rival mechanisms offering a complete alternative philosophy of nature.

Although Weismann and the biometricians adopted different approaches to the study of heredity and variation, it was no accident that their efforts to deal with this outstanding problem coincided in time. After the first flush of enthusiasm, it was inevitable that the hunt for genealogies would bog down in those issues that could not be directly resolved and more emphasis would then be placed on problems associated with the mechanism of selection itself. Weismann's theory merely exaggerated the oppostion of those who found selection unacceptable for scientific or philosophical reasons, but the experimental approach and the rediscovery of Mendelism opened up a powerful second front in the fight against Darwinism. The second front analogy is doubly instructive since, like the powers allied against Germany in the last world war, there was no love lost between the anti-Darwinian forces. In fact, the gradual establishment of Mendelism as the dominant theory of heredity undermined Lamarckism far more effectively than Weismann had been able to do. This left the way clear for the eventual reestablishment of the selection theory once the oversimplifications of the early experimentalists had been recognized. The decades on either side of 1900 were thus the crucial ones for Darwinism, since at that time Lamarckism still flourished in addition to the new Mendelian-mutationist alternative. Darwinism was besieged on all sides, and the fact that the opponents could not agree among themselves could not dispel the general feeling that the inadequacy of the old approach to evolution had now been fully demonstrated.

3

The Decline of Theistic

Evolutionism

 N the early nineteenth century, scientists—at least in the English-speaking world—saw their work as illustrating the divine origin of the universe. Those who made the new discoveries in geology and paleontology did not see themselves as being in conflict with religion. Instead, they restructured their beliefs to accommodate a more sophisticated idea of how God's purpose was manifested in His creation.[1] By the end of the century, this connection between science and religion had been weakened, and even those scientists who professed a religious faith accepted the fact that they must frame their hypotheses without reference to the argument from design. To this extent, the philosophy of naturalism inherent in the Darwinian approach to nature had triumphed. But one should not be misled into exaggerating the rapidity with which science and religion separated themselves in the late Victorian era.[2] Even by the end of the century, Lamarckism was still being promoted because it seemed more compatible with the claim that evolution was initiated by a wise and benevolent Creator. This last remnant of the old tradition had replaced a more widespread belief in design that was still prevalent in the decades immediately after 1859. To many naturalists, Darwin's natural selection was unacceptable as the mechanism of evolution precisely because it left no obvious room for design. They turned instead to theistic evolutionism, the belief that nature unfolds according to a recognizable divine plan.

Theistic evolutionism did not contribute to the eclipse of Darwinism—indeed, it was itself eclipsed by the rise of a naturalistic approach to the question of the origin of species. For this reason I shall limit the depth of my study, although there is certainly room for more detailed work to be done on the topic. I cannot, however, ignore theistic evolutionism, since it anticipated both the attitudes and some of the actual arguments used by the supporters of more naturalistic theories later on. It is impossible to sympathize with the religious views expressed by many neo-Lamarckians without understanding that their position was a genuine attempt to reformulate their beliefs in a manner acceptable to science. The theistic evolutionists had maintained that God's handiwork is directly manifested in the course of nature's development. There were structures and relationships in the living world that could never have arisen by chance or through the blind operation of mechanical laws. These relationships bespoke a purpose underlying the operations of nature, and the laws were merely means by which its goals would be reached. Such a view was in the end recognized as unscientific, in the sense that if the phenomena could *only* be explained as the results of supernatural design, they could not be investigated by science. The next generation of naturalists tried to retain their faith by accepting that nature could design itself, the creative power having been transferred into living matter where it strove both to adapt and to improve the various forms of life. One could no longer prove design in such a system, since the activities of life had to be worked out within a system of law. The Lamarckian explanation of adaptation was, up to a point, just as materialistic as the Darwinian, since it no longer implied that a design was built in to living structures. But the replacement of trial and error by a mechanism in which the organisms' own purposes were achieved allowed one to believe what one could no longer prove: that the whole system was set up by a benevolent God. In this respect, Lamarckism was the direct heir to theistic evolutionism, representing a deliberate attempt to adapt the old argument from design to a more rigorously scientific approach.

The more subtle teleology of Lamarckism replaced the old belief that a fixed state of adaptation proclaimed design by a benevolent external Creator. Yet much of the emphasis placed on design by the theistic evolutionists had necessarily stressed a rather different view of the Creator's purpose. By the very nature of the challenge posed by natural selection, adaptation no longer offered the best evidence of design. In the hope of defining effects that could not be explained by any process of natural development, many theistic evolutionists began to stress relationships between species that had a formal rather than a utilitarian significance. The concept of an orderly, harmonious plan of creation had already been incorporated into the argument from design by naturalists who drew their inspiration from the idealist philosophy of early nineteenth-century Germany.[3] Richard Owen and Louis Agassiz had tried in different ways to stress that the orderly relationships between species were better evidence of design than any number of individual cases of adaptation. Agassiz remained an opponent of evolution, but Owen believed that the pattern might unfold by "natural," that is, lawlike means involving transmutation. This approach became the foundation of most arguments for theistic evolution in the early decades of the Darwinian era. If one could only *hope* that adaptation proved design, one might still *prove* that the orderliness of nature could only be explained in this way.

Theistic evolutionism thus foreshadowed two quite different lines of opposition to Darwinism. On the one hand, Lamarckism allowed the retention of adaptation as a subject for teleology through the more subtle notion of design working itself out within nature. We shall, in fact, see how a number of theistic evolutionists turned explicitly to Lamarckism as the only way of allowing a role for utilitarian factors in their thinking. On the other hand, the formal relationships of the idealist argument from design had a more damaging implication for natural theology. The belief that nature is an orderly system has a fascination all its own, particularly for the student of animal morphology, and that fascination can manifest itself without religious overtones. Indeed, in Germany the idealist philosophy had developed independent of purely Christian concerns. As paleontologists explored the fossil evidence for evolution, they gradually accepted the Darwinian claim that there is no overall goal toward which the development of life is aimed. Order could be found only within the evolution of certain groups, and in many cases the results of these regular developments were far from beneficial. The concept of orthogenesis—regular evolution driven inexorably toward nonadaptive goals—arose when the idealist concern for formal relationships emancipated itself once again from the optimistic attitude of natural theology.

The late nineteenth century was probably the last time that biologists explicitly debated the question of whether form or function is the more basic determinant of organic structure. Darwinism attempted to destroy the belief that life develops according to a formally structured, internally programed pattern. By concentrating on adaptation, it created an image of nature as an unstructured system in which the form of each species follows the demands of function as its lifestyle adjusts to an ever-changing environment. In principle, Lamarckism sided with Darwinism on this issue: It too allowed function to determine form. The idealist argument from design and the theory of orthogenesis both stressed the primacy of pure form, independent of any environmental demand, and thus challenged Darwinism at a far more fundamental level. Yet the fact that both Lamarckism and orthogenesis can be seen to have origins in theistic evolutionism shows that this distinction can easily be blurred in practice. The main concern of all these alternatives to Darwinism was to replace the selection of random variations with some more orderly or purposeful process. One might concentrate on the need for order, and thus stress the formal relationships of orthogenesis, or one might reintroduce purpose through the new teleology of Lamarckism. In theory these two approaches might seem very different, but the fact that both had their origin in a common rejection of Darwinism meant that in practice many naturalists combined them to offer a more comprehensive alternative.

ORDER AND DESIGN

The classic statements of the utilitarian view of design were William Paley's *Natural Theology* of 1802 and the *Bridgewater Treatises* of the 1830s. While they were accumulating endless examples of adaptation to prove design by a benevolent God, however, some European naturalists were taking a very different view. The idealist philosophy that had become popular in Germany revived a Platonic interpretation of nature by treating each species as an element in the overall pattern imposed by Mind upon the material world. Lorenz Oken was perhaps the most extreme advocate of such a view, but in one form or another it was soon taken up by naturalists in Germany and elsewhere. Oken and his followers introduced the law of parallelism, according to which the growth of the human embryo is seen as a direct expression of the hierarchical order of nature and the purposeful, goal-directed character of its operations.[4] Already there were hints that the parallel might be extended into another area: the progressive development of life on earth. Thus emerged the foundations of what would become the "recapitulation" theory, in which the growth of the embryo is supposed to repeat the historical development of life, both processes illustrating the existence of a rational plan underlying all natural developments.

In essence, the recapitulation theory had become popular long before anyone believed that the history of life unfolded through a process of transmutation. Perhaps its leading advocate in the mid-nineteenth century was the Swiss naturalist Louis Agassiz, who remained a convinced opponent of evolution throughout his career.[5] Agassiz had absorbed Oken's *Naturphilosophie* as a student, and although his biological training led him to modify some of the more extravagant claims for the unity of life, he never relinquished the belief that the vertebrates, at least, were linked into a single pattern in which the human form was very important. Agassiz adapted the idealist philosophy of progression to the tradition of natural theology. For him, man was the goal of a divine plan, and was proclaimed as such in a symbolic form by the parallelism between embryology and the sequence of creations. When he moved to America in 1846, Agassiz ensured that his version of the idealist argument from design would become widely known in the English-speaking world. We shall see how the American school of neo-Lamarckism derived its interest in the orderliness of evolution from Agassiz's teaching. Meanwhile, the British anatomist Richard Owen had introduced a somewhat different means of adapting the idealist philosophy to natural theology. His concept of the vertebrate archetype expressed the belief that all forms are modeled on a single basic plan, which has been modified in many different directions to give the species we see in the real world. Owen argued that the unity of type gives a better indication of the Creator's power and wisdom than the individual adaptations of the Paley school.[6] Since Owen's system unified nature through a single underlying form, rather than a single goal of development, it was more easily able to accommodate the idea of a branching process of evolution that would become characteristic of Darwinism.

It was Agassiz's concept of a goal-directed plan of development, however, that was first adapted to construct a scheme of theistic evolutionism. This was exactly the approach taken in Chambers's *Vestiges of the Natural History of Creation,* published anonymously in 1844 and constantly revised until it took final form in 1860.[7] Like Agassiz, Chambers concentrated on the orderly development of life rather than on individual cases of adaptation. He not only saw the progressive order of nature as a divine plan, but insisted that the plan could be programed to unfold without the Creator's direct intervention at every step. A "law" of progressive development was built into nature, ensuring that from time to time new species appeared from old, the new ones always a step further up the hierarchy of organization. The means of transmutation was an extension of embryological growth, allowing Chambers to invoke the recapitulation theory as a necessary consequence of the means by which the Creator's plan is realized. The element of design was absolutely crucial in Chambers's system: The only explanation of why the changes always took place in a progressive direction was that God intended them to. The Creator has programed the law of development into nature in the same way that Charles Babbage could program his computer to undertake a complex sequence of calculations.[8] In effect, the law of progress is merely a predesigned plan whose structure could only be accounted for by invoking the supernatural.

Vestiges was roundly condemned by the scientific community, but it did get the idea of transmutation into circulation. It also served as a prototype for the more sophisticated versions of theistic evolutionism that would emerge in response to Darwinism. Chambers's vision of a linear plan of development was already obsolete by 1859; as Owen recognized, the pattern of development would have to be pictured as a branching one, leading off in many different directions. Nevertheless, the later theistic evolutionists followed Chambers in looking for regular patterns as the best evidence of design. The system that had been rejected as virtually atheistic in 1844 was now revived as a fall-back position by those who wished to preserve the role of design against Darwin's more militantly naturalistic theory. The patterns of development might now have to be more limited in scope, but they were the only hope of preserving some evidence that nature changed in accordance with a divine plan.

We can see how three of the leading British theistic evolutionists made use of the idealist argument from design. St. George Jackson Mivart, the Duke of Argyll, and William B. Carpenter all appealed to the regularity of natural developments as evidence that a rational plan was involved. Mivart's *Genesis of Species* (1871) has already been cited as a leading source of arguments against natural selection. Many of the arguments were directed against the gradual accumulation of minute variations, so that Mivart emerged as one of the early advocates of evolution by discontinuous steps. He suggested that one picture a polygon rolling from one stable position to another, an analogy that was also used by Francis Galton.[9] In addition, Mivart urged as evidence against any purely naturalistic theory of evolution the existence of numerous parallel developments in the history of life. How, for instance, could a cephalopod such as the squid have developed an eye that is very similar to that of man? [10] That two such widely separated lines of evolution should have accidentally acquired the same organ was unthinkable. The parallel must indicate that certain forms have somehow been programed into evolution by the Creator of the system. The steps in evolution may be elicited by an external stimulus, but they are directed by an internal plan built into the very nature of life. In the analogy of the rolling polygon, it is the polygon's fixed shape that determines the next stable position, not the force that tilted it.

Mivart's claim that all significant variation is predetermined in accordance with a divine plan was echoed by the Duke of Argyll. So strong was the element of design in Argyll's *Reign of Law* that Alfred Russel Wallace accused him of advocating supernatural interference in the course of evolution.[11] In later editions of his book, Argyll replied insisting that he meant no such thing: Design was worked out solely through a process of natural law. Although he adopted an idealist view of law in which the Creator's will provides the force linking cause and effect, Argyll was unwilling to allow Him to influence events in an arbitrary manner. The various developments in evolution were all predetermined, having been built into living matter as potentialities at its creation.[12] As evidence for the existence of a preordained plan, Argyll cited the fundamental similarity of all vertebrates as expressed by Owen's archetype.[13]

A more restricted line of evidence for patterned evolution was offered by W. B. Carpenter. In his investigations of the Foraminifera—minute sea creatures that produce delicately structured shells—Carpenter had found geometrical relationships linking the various forms. He inferred that these corresponded to evolutionary sequences, thereby indicating that a particular pattern of development had been built into the group by its Creator.[14] Although Carpenter found all the elements in his sequence among living forms, his evidence closely anticipated that later used by paleontologists to support orthogenesis. The "law of acceleration of growth" proposed by the American paleontologists Edward Drinker Cope and Alpheus Hyatt revived on a smaller scale Chambers's idea that the pattern of evolution unfolds through a step-by-step extension of embryological growth. Both regarded the pattern of development within any group as essentially linear and were seen as leading proponents of orthogenesis. Yet in his first paper, published in 1868, Cope insisted that the pattern of growth was predetermined by the Creator.[15] In this case, the transition from theistic evolutionism based on the idealist concept of design to a theory of pure orthogenesis was quite explicit. The idealist philosophy was even stronger in Hyatt's thought, but the apparent lack of any theological connection allowed him to translate the idea of patterned evolution directly into a theory of degenerative orthogenesis.

VARIETY AND DESIGN

If the idealist philosophy promoted a search for a pattern linking the forms of life together, the question of why evolution branched in so many different directions remained. By 1859, Owen and others had established the branching nature of the history of life, interpreting the divergence as the specialization of many forms for different ways of life. Instead of each species being perfectly adapted to its own lifestyle, as Paley had supposed, there was a steady increase in the level of specialization as the successive forms diverged further from the archetype. Owen himself equated this process with a divinely ordered sequence of transmutations in his theory of "derivation" (1866).[16] Similarly, Argyll argued that the Creator intended to maximize the sheer variety of living forms, although he took the unusual step of suggesting that this was at least in part based on aesthetic considerations. He believed that the brightly colored plumage of the humming birds, for instance, could only be explained by supposing that the Creator designed beauty into evolution for its own sake.[17] For most naturalists, however, the main purpose of divergence was to produce different structures, each with its own way of adapting to the external environment.

The only problem was that a process of adaptation could now be explained quite naturally via Darwin's mechanism of selection. Theistic evolutionists inevitably tried to discredit selection, but their willingness to invoke orderly patterns in evolution shows that they realized the unsatisfactory nature of evidence for design based solely on adaptation. Nevertheless, if one could no longer prove design through adaptation, one could still believe that the adjustment of form to function was a sign of divine benevolence. Argyll, at least, seems to have held that all adaptive trends were built into life as potentialities at its first creation, to be unfolded by a process of law as required. The most subtle discussion of the relationship between adaptation and design came, however, from the American botanist Asa Gray, who was seen as a supporter rather than an opponent of Darwin. Certainly, Gray was one of the few who refused to admit nonadaptive processes in evolution, leading one modern commentator to treat him as a true Darwinian.[18] In principle, he tried to argue that *any* process allowing the species to adapt to its environment is a sign of divine benevolence; but in the end, he seems to have realized that this was a dangerous claim to make on behalf of a mechanism such as natural selection, which seemed to enshrine selfishness as the driving force of nature. Gray suggested that somehow a benevolent God must have directed variation along beneficial lines.[19] Darwin responded by pointing out that this would make selection superfluous and would leave the true course of evolution still in the realm of the supernatural.[20]

Darwin's point was valid, but for those who did want to retain a role for design there were a number of problems to face. Was there, for instance, any real difference between a predesigned process of adaptation and one that merely worked through law? Argyll was the only theistic evolutionist who tried to defend the traditional view that there might be some observable characteristic of the process that would confirm its divine origin. He rejected the Darwinian explanation of vestigial organs, according to which they are the remnants of once useful structures now left to atrophy because they are no longer needed. Instead, Argyll maintained that these were incipient structures being prepared for some future use, and that they would gradually increase in size until they would be able to serve a useful purpose in some new situation.[21] If one could really show that structures evolve to anticipate future requirements, then teleology would indeed be reintroduced in a manner requiring us to believe in an external Designer. The regularities that the scientist calls laws of nature work through cause and effect, which allows the past—but not the future—to govern the present course of events. Perhaps inevitably, given the changing climate of opinion, no one took Argyll's suggestion seriously. If adaptation was to work through law, it would have to be a law based on the principle of causation.

In a sense, there was nothing unreasonable about Argyll's suggestion. If one did believe that the trends leading toward an adaptive goal were built into life by its Creator, there was no reason why He should not anticipate things in this way. Even Gray accepted the idea of pre-designed ends worked out by law when he suggested an analogy with a billiards game: The design of the shot is built in by the impulse the player gives the ball, but it is realized through the blind operations of the laws of mechanics.[22] The lack of attention paid to Argyll's view of rudimentary organs shows that no one was prepared to take this approach to its ultimate conclusion. There was, in any case, another problem. If God has programed life to unfold in certain ways, how did He manage to cram so much information into its original structure? It may have seemed plausible that He built in a simple law of progression, or even a series of linear trends, but how can we believe that all the twists and turns of adaptation required by every species throughout its history were programed into the first ancestors of life on earth? The sheer variety of living structures makes it almost impossible to believe that design could have been imposed in this way. An omniscient Creator might be able to foresee all future requirements, but His ability to build in trends to satisfy these requirements would still be limited by the complexity of organic structures.

The only alternative was a causal mechanism for adjusting each species to its environment in a manner that avoided the harsher implications of natural selection. To be consistent with the basic principle of allowing a scientific investigation of the origin of species, design could not be built into the system from an external source, but would have to become a process operating within the system through the everyday laws of nature. The adjustment of form to function would have to take place not through a predesigned anticipation of future needs, but through a day-by-day interaction between the species and its environment. This was exactly the point made by Samuel Butler when he advocated a new teleology based on such an interaction.[23] It might not allow one to prove design by a supernatural Creator, but it would indicate that life itself is a purposeful force injected into nature. Butler, and many others who came to believe in the ability of life to react positively to any external challenge, held that God could now be seen to act within the universe rather than from outside it.[24] So long as the process of adaptation worked through a purposeful response of the organism to its environment, rather than through trial and error, a new kind of teleology could replace the old argument from design.

The solution was, of course, what we have come to know as Lamarckism. A purposeful adaptation of the species could only be achieved through the cumulative inheritance of characters acquired by the individuals in their dealings with the environment. Even as Butler was advocating the need for a new teleology, the older generation of theistic evolutionists was turning to Lamarckism as a solution to the problem of adaptation. To begin with, however, the influence of Lamarck himself was not recognized, or at least not admitted. Perhaps his reputation had diminished too sharply in the first half of the century for him to serve as a symbol for this hesitant move toward the inheritance of acquired characters. Only later in the century, when the movement had achieved a degree of coherence as an alternative to Darwinism, would the need for a convenient name lead to the rehabilitation of Lamarck's reputation by a generation of naturalists who had been very little influenced by his actual writings.

It was in *Vestiges* that this altered perspective first appeared. Chambers had always been critical of Lamarck for supposing that use-inheritance could account for the progress of life along the chain of being. (This was quite unfair, since Lamarck had postulated an autonomous progressive trend,[25] but like most of his contemporaries, Chambers associated Lamarck's name with only one mechanism of evolution.) Yet, in the later editions of *Vestiges,* he conceded that in order to account for adaptation, the transmutationist would have to invoke "another impulse, connected with the vital forces, tending in the course of generations to modify organic structures in accordance with external circumstances."[26] The 1860 edition of *Vestiges,* although again ostensibly critical of Lamarck, admitted that the direct effects of habit were responsible for creating adaptive structures.[27] Thus, a compromise was reached: The unifying trends that give order to the history of life could be supposed to originate directly from the Creator, but the day-to-day problems of adapting to an ever-changing environment could be left to the purposeful—but quite natural—process of use-inheritance.

A similar slide into Lamarckism can be seen in the writings of Mivart and Carpenter. In the *Genesis of Species*, Mivart had conceded that some of his arguments against selection would not be effective against Herbert Spencer's essentially Lamarckian theory of adaptation.[28] Samuel Butler actually claimed that it was this passage that had led him to investigate and accept the Lamarckian alternative.[29] Over a decade later, in 1884, Mivart himself accepted the Lamarckian approach by conceding that the internal forces he invoked in 1871 did not control the whole of evolution. He now saw that "the characters and the variation of species are due to the combined action of internal and external agencies acting in a direct, positive and constructive manner."[30] A discussion of the effects of habit and a comparison between instinct and heredity (Butler's point) make it clear that the interaction would work through the inheritance of acquired characters. Carpenter too wrote on the hereditary transmission of acquired mental habits and made it clear that no sharp distinction could be drawn between mental habits and the physical changes they produce in the body.[31] Even Argyll took an interest in Lamarckism, although he would not admit it as an adequate alternative to selection.[32]

Confronted with the impossibility of using adaptation as direct evidence of design, several British theistic evolutionists thus accepted Lamarckism as the only way of preserving a role for purpose in nature. This idea thus paralleled Samuel Butler's explicit attempt to connect Lamarckism and teleology. A balance between predesigned orderly trends and the Lamarckian mechanism of adaptation proved the best possible compromise for defending the traditional belief in a divinely ordered universe against the trial-and-error philosophy of Darwinism. We shall see below how the American school of neo-Lamarckism developed perhaps the most influential version of this approach for the explicit purpose of defending natural theology.

LATER DEVELOPMENTS

By the end of the century, the idea of designed evolution had been largely discredited, at least within the scientific community.

Argyll's last attack on Darwinism appeared in 1898, but by that time the theory's supporters were far more worried about the new generation of alternatives. Even outside the realms of science, the philosopher John Dewey could ridicule theistic evolutionism as "design on the installment plan."[33] Perhaps the only remaining approach that a scientist could take to defend the argument from design was that suggested by Lawrence Henderson in 1913.[34] He suggested that the physical structure of the universe appears to be carefully "adapted" to allow the development of life. Even today, it has been argued that if any one of a number of basic physical constants were slightly different from what it actually is, life as we know it would be impossible. As Henderson pointed out, however, this says nothing about the forms of living structure that will evolve. The universe is designed to allow open-ended, not goal-directed, evolution.

Nonscientists, too, were reluctant to revive the traditional notion of design. Yet there were a number of efforts to challenge selection in a spirit reminiscent of the old natural theology. The selfish nature of the "survival of the fittest" still disturbed many people. In science, as I have already noted, Lamarckism provided the best alternative with its more purposeful image of how evolution works, but from outside science, there were still other suggestions. Braving the scorn of the professional scientists, these outsiders sometimes linked their ideas with more orthodox theories, and in so doing they once again help us to understand the potential implications of certain anti-Darwinian positions.

One suggestion that received wide publicity consisted of an attempt to turn the selection mechanism on its head. Not competition but cooperation was the true driving force of evolution: The most successful animals were always those that worked together, and thus evolution had always tended to promote the social instincts. This point was made by the Russian emigré Peter Kropotkin in a series of articles published between 1890 and 1896, subsequently collected in the book *Mutual Aid.* [35] Kropotkin explicitly attacked the pessimistic message of T. H. Huxley's "Evolution and Ethics," which had pointed out the amorality of natural selection. Instead, he offered field evidence to show that animals frequently did cooperate, and suggested that the success of the social instincts could replace selection as the cause of evolution. A more deliberately religious implication was drawn from the same theme in 1894 by the Scottish clergyman Henry Drummond in his *Ascent of Man.* Drummond in particular pointed to the need for parents to look after their offspring as a factor demanding ever greater levels of altruism as evolution progressed. Thus, the divine origin of the process was proclaimed by the enshrinement of love as the driving force in the development of life.

The problem with this argument, then as now, is that the fact of increasing levels of social behavior does not in itself count against selecttion. To those who look only at the emotional level, it may seem that selection and altruism are incompatible, but scientists looking for a mechanism of evolution will soon ask *why* the more social animals replace the less. They may well answer that it is because the more cooperative individuals do better in the struggle for existence than their less sociable neighbors. This is the approach taken to its ultimate conclusion by modern sociobiology. Although our modern techniques were not available at the turn of the century, the few scientists who paid any attention to Kropotkin's and Drummond's work were quick to point out the non sequitur in their argument. The Darwinist F. A. Bather pointed out in a vicious review the failings of Drummond's book as a scientific theory.[36] In a rather more sympathetic tone, Patrick Geddes and J. Arthur Thomson had already suggested that those who looked to parental care as a source of altruism would have to accept that it could only develop in parallel with selfish instincts.[37] In the end, the only way in which the Kropotkin-Drummond thesis could serve as a valid basis for an alternative to Darwinism would be if another mechanism could be suggested for enhancing the social instincts. The only possibility was, of course, Lamarckism: Altruism would be learned by all and then inherited as an instinct. It can be no coincidence that Kropotkin eventually began to write in favor of the inheritance of acquired characters.[38] As with theistic evolutionism, the attempt to challenge selection on moral grounds led straight to Lamarckism as soon as a degree of scientific rigor was demanded.

A very different view of evolution was advanced in 1907 by the French philosopher Henri Bergson in his *Creative Evolution.* Bergson shared the common distrust of natural selection because of its materialistic implications. Like many Lamarckians, he had also recognized that it was no longer possible to imagine evolution as drawn toward a goal imposed by an external Designer. Yet Bergson would not accept the Lamarckians' claim that the consciousness of the individual organisms determines the course of evolution by directing use-inheritance.[39] The nonmaterial factor in evolution would have to be conceived as a basic impulse—the *élan vital*—injected into life at the beginning and constantly trying to express itself by organizing recalcitrant matter into ever higher states. Thus Bergson was more interested in the biological evidence for orthogenesis, and he expressed skepticism about the very limited evidence for the inheritance of externally acquired characters.

The regular, parallel lines of orthogenetic evolution indicated that there was a common impulse to life, which ensured a certain level of harmony in its developments, despite being subdivided into competing branches by its efforts to control the material world.[40] In contrast to the scientific supporters of orthogenesis, who tended to suggest a physicochemical factor controlling variation from within the organism, Bergson proclaimed the existence of what was, in effect, a spiritual force imposing a rational order on the development of life.

The different reactions of Kropotkin and Bergson to Lamarckism make it easier to pinpoint some of the tensions underlying the uneasy alliance of anti-Darwinian theories at the time. The optimistic interpretation of Lamarckism grew out of the old belief that adaptation proved the existence of a benevolent Designer. The Creator's willingness to allow life to design itself by use-inheritance not only allowed one to see nature as a purposeful system, but could also still be taken as a sign of His benevolence. The attempt to present mutual aid or altruism as the driving force of evolution would naturally be linked with this effort to see nature as a system based on a moral foundation. In contrast to Bergson's views as a moral philosopher, his creative evolution took the opposite course of stressing order in nature, thus continuing the idealist tradition of seeking a factor that would unify the bewildering array of living forms. The creative impulse was detectable because of its rational rather than its moral effects. Where the Lamarckians allowed form to be determined by function as a means of giving life the power to design itself, Bergson joined the supporters of orthogenesis in stressing a purely formal control of evolution. Nevertheless, the fact that Bernard Shaw could call his version of Lamarckism "creative evolution," and could try to tell Bergson to his face that this was what he really ought to have meant by the term,[41] suggests that one must take great care when trying to assess the relationship between these two viewpoints. My study of both theistic and creative evolution has suggested that an alliance could be made between Lamarckism and orthogenesis against Darwinism. Because of common origins and common opponents, two quite different approaches to the relationship between form and function could be seen as lending support to each other.

4

Lamarckism

I N his *Philosophie zoologique* (1809), J. B. Lamarck had postulated two causes of transmutation: an inherently progressive trend that forced living things gradually to ascend the chain of being, and the inheritance of acquired characters as a mechanism for adapting them to an ever-changing environment. The Lamarckians of the late nineteenth century concentrated principally on the latter effect, and indeed "Lamarckism" has come to mean little more than a belief in its validity as an evolutionary mechanism. As Stephen Gould has pointed out, however, some Lamarckians— particularly those of the American school—did not altogether ignore Lamarck's other suggestion of an internal directing force.[1] The distinction between characters acquired purely in response to the environment and characters acquired by an environmental stimulus that then elicited an internally-controlled variation trend was not easy to draw in prac-tice. Indeed, it was not always easy to draw in principle, since some Lamarckians were anxious to ensure that, however the new characters were caused, they followed a regular path and thus gave rise to orderly patterns of evolution. Thus, there were two kinds of Lamarckians: those who wished to link the theory with the idea of regular evolution and orthogenesis, and those for whom the inheritance of acquired characters was purely a mechanism of adaptation, more purposeful than Darwinism but no more likely to generate regular patterns of evolution.

Belief in the inheritance of characters acquired during the lifetime of the parent had been widespread in the period before 1859,[2] although few had accepted Lamarck's claim that it was a force powerful enough to change a species. The strongly negative reaction of scientists such as Cuvier and Lyell created an image of Lamarck as someone who had made wildly exaggerated claims about the effectiveness of his mechanism in order to establish a case for transmutation. I have already noted

[58]

that even writers such as Chambers, who came to accept the inheritance of acquired characters, did their best to dissociate themselves from Lamarck's name. Darwin accepted the effect himself, but proposed his alternative of natural selection as a way out of the impasse created by the supposed inadequacy of the Lamarckian mechanism to explain most cases of transmutation. Darwin may have succeeded in converting the scientific world to evolution precisely because he was able to free the theory from the discredited image of Lamarckism; but once evolution itself was accepted, there was no going back. If selection in turn was shown to have major limitations, the inheritance of acquired characters remained as the only alternative mechanism of adaptation, and it was inevitable that naturalists would begin to take another look at it. From the start, some prominent figures associated with the new evolutionism, including Herbert Spencer and Ernst Haeckel, accepted the inheritance of acquired characters as a force at least equal to natural selection. The question of relative strength was not crucial as long as the Darwinians themselves accepted at least some role for the Lamarckian mechanism, but as Weismann's influence grew during the 1880s, a polarization of opinion became inevitable. By the end of that decade a distinct Lamarckian movement had emerged in opposition to neo-Darwinism. The term "neo-Lamarckism" was coined in 1885 by the American scientist Alpheus Packard.[3] By 1899, even E. Ray Lankester could admit that Lamarckism was not a discreditable nickname but a "reputable denomination" for those wishing to establish the inheritance of acquired characters as an alternative to Darwinism.[4]

The popularity of Lamarckism seems to have peaked in the 1890s, but the degree of its success varied from country to country. It was in America that the most coherent school appeared, as even French writers admitted.[5] I shall deal with the American school as a separate movement in a later chapter, but I should note here that there were special circumstances ensuring that American neo-Lamarckism had a unique character with a strong link to orthogenesis. European Lamarckism was more diffuse. In France there were again special circumstances: Darwinism was not the vehicle by which French naturalists were drawn to evolutionism, and it was perhaps inevitable that as the conversion did occur, Lamarck's name would be revived as the symbol for their distrust of the trial-and-error philosophy. In Britain and Germany, the situation was less clear-cut. Because Darwinism retained a significant influence there, the historian's problem is to assess the strength of the Lamarckian opposition. In the late 1880s, Spencer could still lament that most naturalists remained strict Darwinists.[6] By 1890, however,

Samuel Butler—never one to exaggerate the effect of his writing on the scientific community—struck a more optimistic note in pointing out that one could hardly pick up an issue of *Nature* without reading more on the inheritance of acquired characters.[7] If many of the big names remained Darwinists, an increasingly vociferous group of younger naturalists was ensuring that Lamarckism got a hearing. As late as 1910, Peter Kropotkin could write of the division of biology into two camps and could imply that the Lamarckian view had become the more popular.[8] Kropotkin's perceptions, however, were incorrect; it is clear that by this time Lamarckism was beginning to wane due to the growing dissatisfaction of experimental biologists and the success of an alternative theory of heredity. In Germany, a similar pattern seems to have been followed. In the 1890s a vigorous opposition to Weismann was led by Theodor Eimer, while in the period from 1900 to the beginning of the war Paul Kammerer led the experimental defense of Lamarckism, which served as a prelude to the disastrous midwife toad affair of the 1920s.

Support for Lamarckism was generated in a number of different ways. Spencer and Haeckel both erected philosophical systems that took the inheritance of acquired characters for granted—and were then locked into a situation in which the mechanism had to be defended against later challenges. Virtually all Lamarckians were aware of the potential (and potentially contradictory) implications of their theory— either it was a means of preserving religious or moral values, or it promised a more rapid improvement of the human race. There were many scientific arguments as well: In America especially, paleontologists saw Lamarckism as the most logical explanation of the linear trends they saw in the fossil record. Some field naturalists found phenomena that also seemed more easily explicable in Lamarckian terms. In addition, there was a constant fund of support from medical people, animal breeders, and horticulturalists, all of whom could relate stories apparently confirming the inheritance of acquired characters. The crucial question of the 1890s became, Could any of these examples be translated into a laboratory demonstration? Lamarckians certainly tried to rise to this challenge and to adapt their theory to the new wave of experimental biology. Their tragedy was not so much the failure of the experiments as it was the constant ability of their opponents to come up with alternative explanations of the results that bypassed Lamarckism. In the end, the theory was abandoned, not because it lacked proof, but because Mendelian genetics proved so much easier to elaborate into a conceptual foundation for the study of heredity.

One immediate problem was the extent to which acquired characters were supposed to be inherited. Most Lamarckians assumed that however pronounced a new character acquired in a single generation, only a very small proportion of the change could be inherited. Some even proposed that *none* would be inherited until sufficient "pressure" had built up to overcome the species' resistance. If this were so, then no amount of experimental work would reveal proof—although by the same token no experimental disproof would count. As late as 1927, J.B.S. Haldane conceded that there might be a Lamarckian effect working so slowly that it only became apparent over geological time.[9] In this case, Lamarckians could not hope to influence the growth of experimental biology and could hardly complain if the conceptual structure of the experimentalists' theories excluded their effect. They would also have to abandon their claim that Lamarckism was a comparatively rapid mechanism that would give human beings control over their own progress. Obviously, the experimental Lamarckians were convinced that in at least some cases a significant proportion of the acquired character would become hereditary within a few generations. But the very fact that natural evolution is so slow would then suggest that the effect was not usually involved. Lankester was the first to point out that there was also a conceptual difficulty. Lamarck had proposed two laws of evolution: (1) that the individuals of a species can adapt themselves to new conditions, and (2) that these adaptations could be inherited and would thus accumulate. As E. B. Poulton put it in a follow-up to Lankester's argument, "The first law assumes that a past history of indefinite duration is powerless to create a bias by which the present can be controlled; while the second assumes that the brief history of the present can readily raise a bias to control the future."[10]

Of necessity, then, most of the arguments for Lamarckism were of an indirect or circumstantial nature. By the early twentieth century, the Darwinians had become quite adept at pointing out the inadequacy of many of these arguments. J. Arthur Thomson listed a series of misunderstandings upon which the majority of Lamarckian claims were based.[11] It was frequently assumed that if a structure looked as if it had been created by use, then use-inheritance was the most logical explanation of its origin. Many supposed examples of the inheritance of acquired characters were cited without adequate proof, or even without any conception of what might count as adequate proof. In some cases, new genetic characters were mentioned without any effort being made to check whether the character was orginally produced within the lifetime of a single individual. In other cases, genuinely acquired characters were cited without any attempt to determine whether they had truly become inherited, that is, by showing that they were retained by a new generation not exposed to the new conditions.

Perhaps the most significant range of indirect arguments was associated with the belief that individual growth (ontogeny) recapitulates the evolutionary history of the species (phylogeny). Stephen Gould has shown how a Lamarckian view of variation and heredity lay at the foundation of this belief as expressed in both Haeckel's "biogenetic law" and the American school's "law of the acceleration of growth."[12] The Darwinian concept of random variation can best be understood as an assumption that the growth process could be distorted, and that this distortion could occur even in the organism's earliest stages. If, however, the variation is an acquired character, then by definition it is added on during the adult phase; that is, at the end of the growth process. To become inherited, such a character would have to be compressed back into an earlier stage of growth, so that it appeared before adulthood. If this were true, it would then imply that the growth of the organism must pass through a stage corresponding to what used to be the adult form before the character was acquired. This is exactly what is required for an embryo to recapitulate the evolutionary past of the species, and thus it was no accident that the recapitulation theory was taken most seriously by those naturalists who favored Lamarckism. Indeed, the facts of recapitulation were in many cases taken as proof that Lamarckism must be the mechanism of evolution.

What exactly is an acquired character? A good idea of the complexity of Lamarckism can be gained from the fact that it is necessary to distinguish at least four different types. The most obvious example of Lamarckism is the phenomenon known as use-inheritance. In this case, the acquired character is a modification of bodily structure caused by the animal's taking up a new behavior pattern and thus exercising its body in a different way. The parts that are used more will grow in size, just as a blacksmith's arms become more powerful as a result of exercise. The American scientist E. D. Cope called the cumulative inheritance of such functionally acquired characters "kinetogenesis"—the creation of structure by movement. The best example of such a process would be Lamarck's famous example of how the giraffe got its long neck: According to his theory, successive generations of individuals had stretched up to reach the leaves of trees and had thus lengthened their necks. Since the effect was inherited, it had become cumulative over the generations until eventually the modern giraffe had been formed. Use-inheritance serves as the foundation for the claim that Lamarckism is a more purposeful mechanism than selection, since the desires of the animal indirectly control its own evolution. The first giraffes chose a new source of food, after which their behavior and eventually their structure corresponded to the new habit. (Of course this does not mean that an animal can will itself a new structure, but it can alter its behavior and expect its body to respond to the unaccustomed exercise.) Use-inheritance is supposed to be an essentially utilitarian mechanism of evolution, producing only adaptive structures in response to the animals' needs. Note, however, that once fixed, the habit itself becomes an internal directing force controlling evolution independent of further changes in the environment. Although most Lamarckians assumed that the process would be self-regulating, at least one explanation of the orthogenetic production of nonadaptive structures would be the continued effect of a habit beyond the point of utility. Note also that this mode of evolution can only work for animals —something else will be needed to explain plant evolution.

A second kind of character, and one that could be acquired by plants as well as animals, would be the direct, involuntary response any organism can make when exposed to new conditions. An animal raised in a colder climate naturally grows longer hair, while a plant grown in dry conditions might develop a thicker skin to prevent its leaves from losing moisture. Here there is no question of will or effort; the physiological processes that keep the organism alive respond automatically and adjust to the new conditions in which they must operate. Again, if inherited, the new character would gradually be enhanced and would bring about adaptive evolution. Cope called this process "physiogenesis." This kind of Lamarckism would certainly satisfy the natural theologian's belief that evolution has been initiated by a benevolent God. It is not so clear, however, that it would allow one to claim that organisms consciously design themselves. This would obviously not be true for plants or lower animals, where the response would always be purely automatic. One might still argue that for higher animals, the effect would allow them to adjust to a deliberately chosen new environment; for instance, through migration into a new territory.

Both kinds of acquired characters mentioned so far are adaptive because the organism is somehow able to "recognize" what is good for it in the new situation. There must be an internal process that—quite automatically—evaluates what is beneficial and promotes growth in that direction. If a muscle is exercised, it grows; the organism "knows" that this is an appropriate response and is programmed to act accordingly. Such a process will not account for the formation of a character that can only be evaluated by an external relationship. Thus, for instance, physiogenesis could not explain the development of protective coloration, since the internal constitution of an animal's body could hardly recognize the negative benefit of not being seized by a predator. For this reason, those naturalists who studied protective coloration were by far the most convinced supporters of the selection theory. It may be noted that in Steele's modern form of Lamarckism, the process of "somatic selection" plays the role of internal evaluator, thus providing a mechanism for the effect that earlier Lamarckians could only hypothesize.[13]

Why, however, should physiogenesis necessarily be adaptive? Is it not possible that exposure to new conditions might cause, instead of a beneficial response, a distortion of the growth process, which would lead to the creation of a nonadaptive character? Some Lamarckians did, in fact, believe that nonadaptive characters were formed in this way. The advantage of their claim was that it allowed them to use the existence of nonadaptive characters as evidence against selection and in favor of Lamarckism. The disadvantage was, of course, that such a move undermined the more optimistic image of Lamarckism, which rested on the assumption that acquired characters are normally adaptive. Far from opening up the prospect of life being in charge of its own destiny, nonadaptive physiogenesis reduced living things to the status of automata passively reacting to the environment in a totally purposeless manner. Such a view would appear to offer even less hope than selection itself. Indeed, this kind of Lamarckism began to slide rapidly toward orthogenesis. Examples of small-scale nonadaptive evolution stimulated by the environment were actually discussed as evidence for orthogenesis, although, strictly speaking, the term should be reserved for long-range nonadaptive trends. In theory, nonadaptive variation could be attributed to Lamarckism if its direction were determined by the environment, and to orthogenesis if the environment merely acted as a stimulus eliciting internally programmed changes. The line between internal and external control would be extremely difficult to draw, however, and in practice nonadaptive Lamarckism and orthogenesis were seldom clearly distinguished.

In principle, a Lamarckian change must occur in the body of the adult organism before being transferred to the germ plasm. Orthogenesis, on the other hand, originates in a predisposition of the germ plasm to vary in a certain direction, even if an environmental change is required to catalyze the effect. If the new character appears in the earliest stages of growth, this would suggest orthogenesis rather than the inheritance of acquired characters; but again, the distinction breaks down in practice. Even Weismann was prepared to admit a small amount of nonadaptive evolution produced by the direct action of environmental change on the germ plasm, a process in which the adult body did not act as an intermediary. Since the new character originated in the germ, it could not be counted as "acquired," in the normal Lamarckian sense. Yet since Weismann denied any internal predisposition of the germ to vary in a certain direction, his claim that the environment produced the change was frequently taken by Lamarckians as evidence for their own position. In many experimental situations it would be very difficult to decide whether the new character arose in the germ or in the course of the growth process. (The advantage of focusing on adaptive change is that one can then feel reasonably certain that the new character has been "evaluated" by the body before being transferred to the germ.)

The one kind of nonadaptive character that is quite clearly "acquired" is mutilation. Here a totally external force acts on the adult body to eliminate part of its structure. Opinions were divided on whether or not such an artificially acquired character should be inherited according to the Lamarckian theory. After Weismann's famous experiment showed that cutting off the tails of mice did not affect their offspring,[14] most Lamarckians insisted that one could not expect such negative characters to be inherited (although they continued to cite the occasional positive report). Weismann's experiment thus did not stem the tide of Lamarckism, but threw the emphasis onto a search for the inheritance of more positive effects. The most immediate consequence of the experiment was to discredit simple models of heredity such as Darwin's pangenesis, in which each part of the body was supposed to manufacture its own germinal material for transmission to the reproductive organs.[15] Such a theory would certainly allow for the inheritance of acquired characters, but it ought to assume the inheritance of mutilations too, since if a part is missing it cannot manufacture the appropriate germinal material. Lamarckians were thus forced to look for more sophisticated theories of "soft" heredity, in which the reproductive process was integrated into the organism as a whole in such a way that it could not be affected by artificial changes. The difficulty of providing such an alternative theory was to become crucial when Weismann's germ plasm theory was successfully translated into Mendelian genetics druing the early years of the twentieth century. Vague analogies between heredity and memory were not enough to defend Lamarckism against the challenge of the new, experimentally verifiable Mendelian theory.

Enough has been said in this lengthy introduction to show that Lamarckism was not a simple intellectual or scientific movement. Much nonsense has been and still is being written on the subject through a failure to take into account the subtle conceptual issues involved or to examine the implications of all the alternatives. Lamarckism could be made to support the optimistic implications often drawn from it, and there can be little doubt that most of its supporters were consciously trying to find an alternative to the Darwinian philosophy of nature. We cannot, however, afford to ignore the darker side of Lamarckism's implications, as illustrated by the link with orthogenesis and non-adaptive evolution. Even when the adaptive effects were stressed, we shall see that Lamarckism could serve as the basis for policies of both laissez-faire individualism and collectivist reform. A biological theory with this degree of complexity is not a tool with a single philosophical or social application. It is a complex intellectual framework that can be extended in many different directions under the influence of both scientific and external pressures.

THE ORIGINS OF LAMARCKISM

The role of Lamarckism in Darwin's own thought has been discussed often enough. Certainly, he accepted the inheritance of acquired characters, and he seems to have placed a somewhat greater emphasis on that mechanism as he became aware of the antiselectionist arguments. His theory of pangenesis was certainly consistent with the inheritance of acquired characters, although it would be going too far to claim that it was developed principally as a means of opening up this line of escape.[16] A favorite tactic of later Lamarckians was to point out just how strong Darwin's leanings in this direction had been.[17] An entire chapter of the *Variation of Animals and Plants under Domestication* was given over to the direct effect of the conditions of life in shaping variation.[18] Here Darwin listed many examples derived from the experiences of animal breeders to suggest that the Lamarckian principle had been at work. Such examples, although often accepted uncritically, were to become the stock in trade of the Lamarckian movement. Nor did this line of support diminish toward the end of the century, when the great American horticulturalist Luther Burbank would declare: "Acquired characters are inherited or I know nothing of plant life."[19]

Another source of anecdotal evidence was the medical profession. There was a widespread belief that diseases, or at least their debilitating consequences, could be inherited. In many cases, the possibility that the effect might be transmitted by contamination of the growing fetus in utero, instead of genetically, was not seen as invalidating the Lamarckian conclusion. It has been argued that physicians did not at first make any distinction between what we call heredity and the transmission of an effect from mother to offspring during the growth process—both were considered to be aspects of a much more broadly defined notion of heredity.[20] Experimental testing at this level had already by the 1860s given rise to what was to become the classic "proof" of Lamarckism: C. E. Brown-Séquard's work on the inheritance of epilepsy in guinea pigs.[21] Here the epilepsy was induced by brain lesions, and appeared to be inherited, although the neo-Darwinists could argue that no proof was offered that the disease organisms were not transmitted independent of the germ plasm.

In Germany, there was some resistance to Lamarckism because of a legacy inherited from the old transcendentalism.[22] Use-inheritance, no less than Darwinism, subordinated form to function, thereby violating the deep-seated belief that there must be laws of pure form that control the development of living structures. Nevertheless, there were some who adopted a more utilitarian framework, and who saw the direct action of the environment in conjunction with selection as the driving force of evolution. Karl Semper wrote his *Natural Conditions of Existence* in 1880 in order to reject the belief that organs have been formed "in accordance with some transcendental law (or plan) of development."[23] He repeatedly stressed the extent to which structures are adapted to the demands of the environment. The book was hailed as a pioneering work by some later neo-Lamarckians, and it is clear that Semper did accept the direct action of the environment. Yet it is also clear that he gave selection a prominent role too, and that many of his examples of adaptation were uncritically adopted by Lamarckians as evidence for their own position.

In both Germany and Britain, the principal lines of support for Lamarckism originated from a broader view of the philosophy of evolution. The differences between the perspectives of Ernst Haeckel and Herbert Spencer are obvious enough, but there are also some striking similarities that originated in their common support for the inheritance of acquired characters as a major addition to selection. Both saw the Lamarckian effect as playing a significant role in a process of universal evolution whose basic trends were toward progress and diversity of form. Both attempted highly speculative theories of heredity based on the assumption that unknown forces allowed organic matter to absorb impressions from its environment. And yet it is difficult to imagine two more divergent viewpoints concerning the implications of evolution for human beings. Although Spencer remained the apostle of laissez-faire individualism, Haeckel promoted a collectivist social philosophy that has been seen as a stimulus to the later emergence of Nazism.[24]

Lamarck's importance is proclaimed in the subtitle of Haeckel's first popular work on evolution, the *History of Creation*. The text also presents Lamarck as the first to work out a consistently materialistic theory of evolution, despite his ignorance of selection.[25] Nevertheless, when Haeckel offered his own support for Lamarckism he was often confused, on the one hand accepting "sports of nature" as acquired characters and on the other proclaiming a "law of adaptive or acquired transmission."[26] The belief that acquired characters are normally adaptive was a part of Haeckel's optimistic philosophy of progressive evolution, in which Lamarckism was the primary force generating new characters to be tested by selection at the level of interspecies competition. The concept of variation by addition to growth was crucial in Haeckel's commitment to the recapitulation theory, or "biogenetic law," as expressed in his *Evolution of Man*. For all his talk of adaptation, Haeckel retained the idealists' view of evolution as a succession of upward steps toward man, each step having been preserved in the form of a lower animal. Although never a supporter of orthogenesis, Haeckel's concept of the hierarchical development of life as a whole anticipated the more restricted versions of linear evolution proposed by the American school of paleontologists. The embryological analogy almost always promoted a teleological view of evolution, even among those who accepted the need for a natural (that is, Lamarckian) explanation of the successive additions to growth.

The superficial nature of Haeckel's materialism was already apparent in his attempt to create a theory of heredity that would permit Lamarckism: He unveiled it in 1876 in his pamphlet on the *Perigenesis of the Plastidule*.[27] Following an idea first put forward by Ewald Hering,[28] he drew an analogy between heredity and memory and tried to explain the similarity by postulating wavelike motions in the "plastidules," or basic units of living matter. It is debatable whether this was really a materialist explanation of how life could respond to its environment, or whether the memory analogy was to be taken more seriously as an indication that even the most basic form of living matter was aware of its surroundings in the psychological sense. Whatever Haeckel's protestations of the need for purely natural explanations, statements such as "heredity is the memory of the plastidule, while variability is its comprehension" tended to imply a more vitalist interpretation.[29] In the hands of Samuel Butler, the memory analogy was to become the foundation of a near-vitalist interpretation of Lamarckism. The ability of living things to "remember" experiences by incorporating them into their own structures became the driving force of purposeful evolution through the inheritance of acquired characters. It was also a necessary foundation for the belief that in the course of its growth the embryo "remembers" all the stages through which its ancestors have evolved.

The confusion between materialism and vitalism lay at the heart of the philosophy of monism expounded in Haeckel's *Riddle of the Universe*. In this work, matter and spirit were treated as parallel manifestations of one universal substance. Indeed, Haeckel virtually equated force with mind, adopting the pantheist view that God represents the driving energy of the world.[30] Again, this sounds very similar to the view expressed by Samuel Butler and those Lamarckians who saw their theory as a description of how God allowed life to design itself. But unlike these Lamarckians, Haeckel would not allow life the power of free creativity in its attempts to deal with the challenges of the environment. He repudiated the idea of freedom of the will and held that every act is predetermined by the constitution of the organism and the state of its environment.[31] He also denied ever having attributed consciousness to the basic elements of living matter. The plastidules did possess psychic characters of sensation, will, and memory, but these were all expressed unconsciously. Only in man and the higher animals did the development of the brain permit the emergence of consciousness. Thus, Haeckel anticipated the quasi-vitalist viewpoint of many later Lamarckians without developing any of its antideterminist implications. For him, the response of the organism to its environment was essentially predetermined, almost mechanical, and did not represent the expression of an innate creativity in the life force. Progress was the result of basic laws built into matter, not of the creative insights of generations of living things.

In a very different way, the philosopher Herbert Spencer also anticipated many of the arguments that would be used by later neo-Lamarckians, but again within the framework of an essentially deterministic view of progress. Spencer had already been attracted to the Lamarckian hypothesis when he wrote "The Development Hypothesis" in 1851.[32] Darwin's alternative of natural selection made him reexamine his own assumptions, but the *Principles of Biology* (1864) still presented the inheritance of acquired characters as the more significant mechanism of evolution. The deductive style of Spencer's reasoning allowed him to pioneer many of the indirect arguments for Lamarckism. He saw adaptation as a process by which the organism was driven to reestablish an equilibrium with the changing environment through the action and reaction of forces. To deny that such a process could have a permanent effect through the inheritance of acquired characters was to deny the principle of the conservation of energy (in Spencer's terms, the persistence of force).[33] He suggested that heredity was controlled by "physiological units" with an unknown polarizing force capable of responding to external conditions.[34] Exposure to new conditions would produce both useful and random variation in a species, so that both the Lamarckian and Darwinian mechanisms would be able to operate. The one produced a "direct," the other an "indirect," equilibration between the organism and its environment. Spencer conceded that characters depending for their usefulness upon very rare occurrences (for example, attacks from predators) would have to be developed by selection alone. Still, he insisted that "adaptive change of function is the primary and ever-acting cause of . . . variation."[35]

To back up his logical arguments, Spencer cited many examples of effects that he believed could only be explained by Lamarckism. The reduction in the size of the wings in domestic ducks and hens was quoted from Darwin as an illustration of the inherited effects of disuse. The blindness of many cave animals was also attributed to disuse of the eyes—a favorite example of many later Lamarckians. The development of the musical faculty in man, along with the frequency of cases in which musical talent was inherited, were both attributed to use-inheritance. Brown-Séquard's guinea pigs were also mentioned as experimental proof of the inherited effects of disease.[36]

Spencer seems to have made no clear distinction between use-inheritance and nonvolitional responses of the organism to its environment. For him, both were equally inevitable consequences of the organism's natural tendency to regain equilibrium. Both were necessary, almost mechanical, processes of adaptation. Spencer was anything but a precursor of those Lamarckians who saw will or volition as an essential part of the system's creativity. He specifically criticized Lamarck for seeing "desires" as the driving force of the animals' activities, on the grounds that this ignored the primary question of how the desires originated in the breakdown of the original equilibrium.[37] Although Lankester held that Haeckel's attribution of memory to his plastidules was equivalent to the polarizing forces of Spencer's physiological units,[38] it is clear that Spencer was never tempted by the pseudo-mysticism of the identification of spirit and matter.

Although in general Spencer saw evolution as a very slow affair, he accepted the fact that Lamarckism as a theory of human evolution offered the prospect of more rapid advance.[39] Again, however, we cannot see this as an anticipation of the optimistic belief expressed by many later Lamarckians in the power of human beings to direct their own evolution rapidly toward a chosen goal. Such optimism was founded on the idea that volition is the driving force of creative evolution, allowing individuals the freedom to choose their own goals. For Spencer, Lamarckism simply offered a more rapid way of bringing individuals into equilibrium with the changing social situation. There was no question of people controlling their own evolution, since the forces at work were so complex that they could only be left to nature. Spencer felt that individuals must learn how to adapt to their social environment, and they would then teach their children the benefits of their experience. A policy of laissez faire was essential because it was only the constant threat of misery that would keep people up to the mark in an ever-developing society. Free enterprise did not eliminate the unfit—it forced everyone to acquire fitness. Thus, although Spencer subordinated the individual to the impersonal forces of nature, he did so via a Lamarckian, not a Darwinian, mechanism. Like so many of his contemporaries, he failed to see that the "inheritance" of new behavior patterns through learning had nothing to do with the process of biological heredity. The all-too-prevalent assumption that cultural development must be a simple continuation of biological evolution would lead many later Lamarckians to make the same mistake in the context of their own quite different interpretation of progress.

By 1886, Spencer was already becoming worried by the increasingly dogmatic stand of some Darwinians, and he published a new defense of Lamarckism.[40] Many of his original arguments were repeated, along with a reply to the criticism that there was no experimental evidence for the inheritance of acquired characters. He pointed out that the effects of artificial selection were obvious, since in order to be selected in the first place a character had to be visible. In contrast, the accumulation of an acquired character through heredity might be so gradual that the effect could not be easily detected. In 1893 Spencer was again provoked into writing an article under the title "The Inadequacy of Natural Selection," sparking a public debate with Weismann.[41] Spencer distrusted the concept of an isolated germ plasm, which violated all his own principles about the integrated nature of the bodily functions. His statement that "either there has been the inheritance of acquired characters or there has been no evolution" certainly helped to rally support for the cause,[42] but there is also some evidence that Spencer was having second thoughts about the prevalence of the Lamarckian effect. One of his arguments now conceded that use-inheritance could lead to the creation of a trivially nonadaptive factor—the tactile sensitivity of the nose and the interior of the mouth.[43] He also conceded that natural selection operated predominantly among the plants and lower animals, while only in the higher animals did Lamarckism become "an important, if not the chief cause of evolution."[44]

It is curious that Spencer began to concede a more prominent role for selection just as neo-Lamarckism was beginning to crystallize as a self-conscious movement opposed to Darwinism. Nevertheless, his arguments for the inheritance of acquired characters were the foundation upon which the new movement would build. Even opponents such as J. Arthur Thomson agreed that the effect would have to be discussed seriously because it was accepted by a thinker of Spencer's caliber.[45] Haeckel also came to support Eimer and the German opponents of Weismann in the later editions of his popular writings.[46] In their very different ways, these two thinkers had helped to establish certain basic elements of the Lamarckian movement. Spencer's conceptual arguments and Haeckel's link between Lamarckism and the recapitulation theory not only shaped later Lamarckian thought, but also defined some of the tensions within the movement. There was, however, one aspect of neo-Lamarckism that neither Haeckel nor Spencer had been able to anticipate, because each had retained an essentially mechanistic image of how the organism responded to its environment. The new teleology promoted by many Lamarckians grew out of a belief that living things had the freedom to shape their own evolution. Yet this too had been anticipated in the early years of the Darwinian debate, although by a writer who had far less direct influence on the scientific community: Samuel Butler.

Butler claimed to be writing entertainment rather than science, and he had no hesitation in lambasting the professionals as the founders of a new dogmatism.[47] It is thus hardly surprising that his writings on evolution were at first either ignored or treated as a joke. Despite all his extravagances, though, Butler did help to define certain key implications of Lamarckism that were to become widely influential at the end of the century. Although he had no illusions about his influence on the scientific community, we know that he was eventually taken seriously by at least some professional biologists, including Francis Darwin.[48] In thus having his reputation rescued at the very end of his career, Butler symbolized the emergence of a new Lamarckism dedicated to ideals that were derived from his way of thinking rather than from the more mechanistic approach of Haeckel and Spencer.

Although Butler was in New Zealand when the *Origin of Species* appeared, he was soon an enthusiastic convert to Darwinism. His novel *Erewhon,* in which machines were presented as extensions of man's physical body, was seen as a parody of Darwinism, although Butler claimed that he was merely working out the theory's implications. He became concerned, however, about the origin of instinctive habits such as the newborn baby's ability to breathe, and wrote *Life and Habit* to argue that intelligent actions can eventually become instinctive and hence inherited. In effect, Butler claimed that for a habit to be learned perfectly, it must become an unconscious instinct. From psychologists such as Ribot and Carpenter he derived the idea that the body itself can behave intelligently without the brain, through the inherited effects of instinct.[49] From this emerged the analogy between heredity and memory: The body remembers its past activities in the form not only of instincts, but also of the physical modifications the instinctive behavior has produced. Butler claimed that our sense of individual existence is in fact an illusion, since the individual organism is merely an integral part of the species and its evolutionary history.

The memory analogy thus led Butler into an essentially Lamarckian position. He argued that the idea would explain all of the facts of heredity, including of course the inheritance of acquired characters and the recapitulation of past evolution by the growing embryo.[50] Much of *Life and Habit,* however, was written before the crucial point at which Butler realized that his thinking was now turning toward a completely anti-Darwinian theory of evolution. The catalyst was his reading of Mivart's *Genesis of Species,* which first allowed him to see that the basic idea of evolution could be separated from Darwinism, leaving him free to look for a more purposeful mechanism.[51] He was led—apparently by Mivart's reference to Spencer's Lamarckism—to read Lamarck and recognize that here was the foundation of a complete alternative to selection that was more consistent with his own thinking. The last chapters of *Life and Habit* were then written to bring out the possibility that Lamarckism might solve all the conceptual problems raised by Darwinism.

Already, in *Life and Habit,* Butler had insisted that a species changes essentially because it wants to change.[52] This was to become the basis for his reintroduction of teleology into evolution theory in *Evolution, Old and New* (1879). Much of this book is taken up with the assault on Darwin for (as Butler alleged) deliberately concealing the extent to which earlier naturalists such as Buffon, Erasmus Darwin, and Lamarck had developed a full-fledged theory of evolution.[53] This came later; the early chapters are more concerned with the new teleology, in which design comes not from an external Creator but from the ability of living things to fulfill their own desires and thus shape their evolution through use-inheritance. Butler argued that in rejecting the static teleology of Paley, Darwinists such as Haeckel had gone too far in adopting a totally mechanistic view of life. The failure of selection allowed one to reintroduce teleology via Lamarckism, and to see God operating within life instead of shaping it from without.[54] There would, admittedly, no longer be any long-range goal toward which evolution could be aimed. Each step would be taken in response to the immediate problems facing the species, and would thus often fall short; but in the long run, the cumulative effect of such purposeful reactions would be to produce major evolutionary developments. Instead of simply being passive machines, living things had the power to direct their own evolution because their bodies would respond to their activities.

Butler's attack on Darwin continued in *Unconscious Memory* the following year. He now expanded the analogy between heredity and memory by invoking the view of Ewald Hering of Vienna, who had first suggested the idea in 1870. Butler had not been aware of Hering's work when he wrote *Life and Habit,* but he now redressed the balance by providing a complete English translation of Hering's paper "On Memory as a Universal Function of Organized Matter."[55] Like Haeckel, Hering had suggested that the memory process might work through the vibrations of the ultimate particles of living matter. Butler now took up this theme, which—as he later admitted—was not Hering's real interest.[56] Yet it is clear that he did not want to develop a materialistic theory of the properties of life. In effect, he now held that energy is the vivifying power of the universe, so that all matter is, in a sense, alive.[57] This would help to explain the spontaneous generation of life from which evolution began, when there was supposedly no living matter.

Butler's last book on the subject, *Luck or Cunning,* summed up his system of teleological Lamarckism. The title itself was meant to suggest the essential difference between the trial-and-error process of selection and the purposeful activities of life given prominence in Lamarckism. Once again, the Darwinians were accused of trying to eliminate mind from nature to create a philosophy of soulless materialism: "The theory that luck is the main means of organic modification is the most absolute denial of God which it is possible for the human mind to conceive—while the view that God is in all His creatures, He in them and they in Him, is only expressed in other words by declaring that the main means of organic modification is, not luck, but cunning."[58]

Finally, in 1890 Butler wrote "The Deadlock in Darwinism" in a more optimistic mood, proclaiming at last the wider recognition of the views he had been advocating for over a decade. Once again, God was seen as the soul of the universe, working within it as its vivifying power. In response to Weismann's neo-Darwinism, Butler now came out with his most impassioned rhetoric: "To state this doctrine is to arouse instinctive loathing; it is my fortunate task to maintain that such a nightmare of waste and death is as baseless as it is repulsive."[59] Few later Lamarckians, except perhaps Bernard Shaw, would have put it quite so bluntly, but few can have been unaware of the implications that Butler linked with their theory.

LAMARCKISM, 1890–1914

By 1890, the controversy over Weismann's theory was well under way.[60] Not all who opposed the theory were active Lamarckians, but the desire to preserve a role for the inheritance of acquired characters was probably the leading source of opposition. Weismann's name certainly figured prominently in the writings of the increasingly vociferous group of Lamarckians that emerged in the 1890s, as a symbol of the dogmatic selectionism to which they were opposed. To them, the concept of the germ plasm seemed to proclaim the sterility of Darwinism: That the selection theory should have to resort to such an artificial subsidiary hypothesis illustrated the insecurity of its foundations. To all of those who had harbored doubts about the efficacy of natural selection, Weismann's theory served not to uphold Darwinism but to undermine it. In their determination to attack dogmatic neo-Darwinism, they became more ready to promote the Lamarckian alternative that had so long played a subsidiary role. Some, like Spencer, merely wished to preserve a place for use-inheritance within a system that still included selection, but others were now emboldened to explore the possibility that Lamarckism might serve as a complete alternative theory. Once again the claim was heard that selection could be no more than a secondary or negative factor: The primary factor in evolution was the origin of variation, and if once it could be shown that this could be purposefully controlled, the elimination of the less successful new characters would be of little importance.

The 1890s were the heyday of the new Lamarckism. So strong were the indirect arguments in its favor that for a while the paucity of experimental evidence was not considered crucial. To the Darwinians, of course, the weakness of the Lamarckian arguments was obvious from the start,[61] but their objections were at first overshadowed by the wave of support for the new theory. Only after 1900 did the search for experimental evidence become crucial in the eyes of most observers, to the detriment of the Lamarckian position. As Mendelism began to capture the imagination of more biologists, it became widely accepted that the experimental approach could yield hard information on the question of variation and heredity. Whatever the strengths of their indirect arguments, the Lamarckians would have to compete in the same arena or face growing public indifference. Try they most certainly did: Peter Kropotkin reported that he had read over 200 experimental reports published between 1906 and 1909, not including many American contributions.[62] It is thus evident that well-known figures such as Paul Kammerer did not stand alone in their endeavors. At first, the Lamarckians themselves thought they would still be victorious; but the Darwinians and Mendelians continued to expose the limitations of the experiments, showing how they could be explained without invoking a Lamarckian factor. More and more biologists became dissatisfied with wrangling over the "interminable question," and complained that if the Lamarckians could not come up with anything more definite their concept should be shelved, if not abandoned altogether.[63] By the time the war threw European science into a turmoil, biologists had already begun to move away from Lamarckism. Yet the theory was never abandoned completely: Rather than refutation, it faced a gradual loss of interest as the new science of genetics began to show the achievements that were possible with an alternative concept of heredity.

The Lamarckians' greatest weakness lay not so much in the failure of their experiments as in their inability to develop their ideas into a coherent alternative to Mendelism that would serve as the theoretical framework for the laboratory study of heredity. Increasingly they adopted Weismann's—and later the Mendelians'—terminology, and tried only to show that the germ plasm *could* be affected by its environment. The analogy between heredity and memory was their only major alternative, but even this required the postulation of something that would store the information gained by experience. Efforts to define this "something" were unsuccessful, and all too often the Lamarckians fell into the trap of assuming that the information would be carried by a germ plasm similar to Weismann's in all respects except its complete isolation. After 1900 they even tried to claim that their newly acquired characters would obey Mendel's laws. The penalty for this was that Lamarckism inevitably became relegated to a subsidiary position: Even if it were valid, it would represent only a minor addition to the orthodox view of heredity. Thus, the Lamarckians were betrayed by their growing willingness to admit that acquired characters would be inherited through incorporation into a definite substance of heredity— a substance whose basic characters were already being defined by their opponents.

The source of this weakness lay in the Lamarckians' traditional fascination with the growth of the individual organism. Their position demanded that they provide experimental demonstrations of two effects: the adaptation of the growth process to new conditions, and the inheritance of that adaptation. Perhaps inevitably, given the interest in embryology fostered by the recapitulation theory, most Lamarckians chose to concentrate on the experimental study of the forces affecting growth—the new science that the Germans called *Entwickelungs-mechanik*. This was certainly a legitimate branch of experimental biology, but it was a fatal choice for the Lamarckians because it left them isolated from the main thrust of research into heredity, which was thus taken over by genetics. Kammerer might fill the *Archiv für Ent-wickelungsmechanik* with reports of modifications to growth, and even their inheritance, but without a theory of inheritance to back it up this work was of limited value. Evolutionists' attention was increasingly focused on the problem of heredity, where the Mendelian concept of fixed genetic units was proving so fruitful as a guide to experimental study. The experimental Lamarckians had gone off in a different direction altogether, and in the eyes of most biologists had evaded detailed study of what was really the most crucial problem facing their theory. Once they began to use their opponents' terminology to discuss heredity, it was inevitable that their theory would not survive, whatever the success or failure of their experimental results.

This fatal misdirection of the Lamarckians' experimental efforts arose not only from the traditional link between their theory and embryology, but also from a conceptual problem lying at the heart of their view of evolution. There was a basic incompatibility between the belief that the final stages of growth could respond to new conditions in a purposeful way, and the need to see the results of this activity permanently fixed into a hereditable program affecting all future growth. This was exactly what Poulton had meant when he declared that Lamarck's two laws were incompatible: One cannot assume that the germ plasm is vulnerable to external influence and that it will also preserve that influence in a rigidly hereditable form. Extreme hereditarians of the Weismann school were always willing to claim that the true source of any new character lay within the germ plasm. They frequently argued that what the Lamarckians called an acquired character was only an illusion.[64] In a sense, all characters are acquired in response to environmental stimuli, since the germ plasm cannot produce a new organism unless it is allowed to develop under suitable conditions. If a new character appears when the organism is grown under new conditions, this can more consistently be explained by assuming that the potential for it already existed within the germ plasm, and that it was only waiting for the appropriate stimulus. New germinal material would be produced, not by an influence from outside, but by a purely germinal process such as mutation. Such an approach at least had the virtue of explaining apparently acquired characters without assuming that the germ plasm is sometimes absolutely fixed and at other times open to external influence.

The Lamarckians claimed that a purely germinal source of variation was unthinkable. The germ plasm was an integral part of the body, which was in turn exposed to the external environment. One Lamarckian even claimed that the production of new characters within the germ alone would be equivalent to spontaneous generation.[65] An external influence must act upon the germ plasm in order to produce any change in its constitution, but to establish the need for such an external stimulus was not enough, since it could still be held that the stimulus acted directly upon the germ plasm instead of through a new character acquired by the body. We shall see that much of the Lamarckians' experimental evidence could be explained in this way. Some tried to argue that the distinction was of little importance, but in fact it turned out to be crucial. If the stimulus acted directly on the germ plasm, there was no need for the variation to be adaptive, since it had not been evaluated by any bodily mechanism. It would also be easier to see how such a change could rapidly take place under laboratory conditions simply by catalyzing the rearrangement of a germinal element (a gene, in the Mendelians' terminology). Lamarckism would thus reduce itself to nothing more than the environmental stimulation of nonadaptive mutations, at best only a minor addition to orthodox genetics. The fact that many laboratory experiments did involve sudden changes, and that efforts were made to show that the new characters bred true according to the Mendelian laws, suggests that some Lamarckians were falling into this trap.

The true Lamarckian position depended upon the claim that the body, when exposed to new conditions, can acquire a character for which there is at first no element within the germ plasm. In order for the character to become inherited, however, such an element would gradually have to be created, and it was difficult to see how this could occur if all characters are supposed to be the results of distinct units in the germ plasm. The only hope of formulating a truly environmentalist theory was to abandon the whole notion of a hereditary substance in which all the information needed to produce a new organism is permanently encoded. This had been the implication of Darwin's pangenesis, but it was no longer possible to believe that the body manufactures its own hereditary material in each generation. The remaining alternative was to assume that the purposeful activity of growth, which forms an adaptive character in response to an environmental stimulus, is typical of the process of ontogeny. Instead of seeing the organism as completely preformed within the germ plasm, one could imagine growth to be a totally open-ended process, shaped at each point by the forces prevailing within the egg or womb. The character of the new organism would thus be determined by the nature of the conditions created by the parents for its growth. If the external environment forced the parents' bodies to change, the resulting acquired character would naturally be passed on through an alteration in the conditions of the growth process.

Such a radically environmentalist approach to heredity was actually suggested by a few extreme Lamarckians, including the American scientist John A. Ryder and the British paleontologist Arthur Dendy. It was also hinted at in Yves Delage's classic survey of heredity. Ryder freely admitted that this perspective totally blurred the distinction between acquired and hereditary characters,[66] while Dendy conceded that the newly acquired characters would only continue to be produced if the changed conditions were maintained.[67] As he pointed out, that is exactly how evolution works: If the climate changes, the change is irreversible and will exert a permanent effect on the species. Such a view necessarily implied, however, that the experimental search for acquired characters that continued to be inherited after the new conditions were withdrawn might turn out to be futile, or at best would have to be confirmed to nonadaptive characters. Dendy suggested that the environmentalist view of heredity uncovered the true cause of progressive evolution, and he relegated the Mendelian effects to a secondary, trivial level of heredity that was indeed controlled by the material structure of the germ plasm.[68] In the words of Delage, who acknowledged that external action on the germ plasm stimulated only nonadaptive variation, "Phylogeny creates organs without regard to function: ontogeny takes these organs and, as far as possible, adapts them to necessary functions."[69]

Dendy's support for an environmentalist interpretation of ontogeny was part of a wider plea that sciences such as paleontology should not be hindered by the small-scale discoveries of the laboratory biologists. Like his better-known American counterpart, Henry Fairfield Osborn, he believed that fossil trends reveal phenomena that might never be duplicated in the short time scale of a laboratory experiment. At this point, the debate over Lamarckism degenerated into a dispute over priorities between different areas of science that could not agree on a common theoretical postion. The central problem of Lamarckism was that the environmentalist insights gained from paleontology and from field studies of organisms adapting to new environments could not be translated into a theory applicable to laboratory studies. To the Mendelians, obsessed with the rigid inheritance of unit characters, Dendy's position must have seemed like a reductio ad absurdum of the whole idea of environmental influence. It cannot be denied that their own extreme hereditarian philosophy was equally shortsighted, but its advantage was that it was immediately applicable to the increasingly poplar laboratory approach, while the environmentalist theory seemed to ignore those phenomena that were most easily demonstrable. For this reason, most Lamarckians chose to ignore the environmentalist theory of ontogeny. They tried instead to adapt their theory to the concept of fixed hereditary units, and in so doing condemned themselves to a secondary role.

Although Lamarckism proved impossible to transform into a complete theory of heredity, its basic tenet held that in the later stages of growth the organism does have the capacity to respond in a purposeful manner to changed conditions. Instead of a mechanical, trial-and-error process of selection, adaptation resulted from the active powers of living organisms. For this reason, Hans Driesch admitted that Lamarckism promoted a vitalist philosophy, as opposed to the materialism of the selection theory.[70] Yet the link between Lamarckism and vitalism should not be overstressed, since we have seen that the inheritance of acquired characters could be supported by thinkers as diverse as Herbert Spencer and Samuel Butler. There were some attempts to reduce the purposeful, evaluative function of the body to a selective and hence purely mechanical process. Driesch himself attacked a suggestion by August Pauly that the animal merely makes use of those random variations that happen to fit its needs.[71] A process strongly resembling Steele's modern idea of somatic selection was proposed by the materialist philosopher Félix Le Dantec as a basis for Lamarckism.[72] Even those Lamarckians who accepted the innate responsiveness of living matter were often at pains to insist that they saw the process operating in a purely mechanical fashion. Thus, Richard Semon's development of the memory-heredity analogy was based on the assumption that "memory" is merely a shorthand description of a material system of information storage.[73] The very fact that Driesch, the most prominent vitalist of his day, was only a lukewarm advocate of Lamarckism should make contemporary historians understand that the inheritance of acquired characters was an effect so basic that it could be incorporated into any philosophy of life, vitalist or materialist.

One should be even more careful in interpreting the link between Lamarckism and the claim that consciousness is the creative force in animal evolution. Certainly there were some Lamarckians who insisted that use-inheritance allowed evolution to follow a path mapped out by the consciously chosen reaction of the animal to its environment. The American paleontologist Edward Drinker Cope was the best-known scientific advocate of this view. Indeed, there is some evidence that the term "neo-Lamarckism" was reserved for this interpretation, with "Lamarckism" sufficing for a less extravagant belief in the inheritance of acquired characters.[74] The growing popularity of Butler's works may have helped to promote the link between Lamarckism and mental creativity outside the scientific community. Its importance as a counter-argument against materialism can be gauged from the fact that some Darwinians tried to insist that their theory did not deny the role of purposeful behavior in evolution.[75] The best illustration of this is the mechanism of "organic selection," sometimes called the "Baldwin effect" after its American co-discoverer, James Mark Baldwin. In Britain the leading advocate of this idea was the psychologist C. Lloyd Morgan.[76] The mechanism was based on the assumption that a deliberately chosen new behavior pattern could influence the evolution of a species, but not by means of the inheritance of acquired characters. The body of each animal would adapt itself to the new situation, but instead of this adaptation being directly inherited, it would give the species time during which random variation could come up with truly hereditable equivalents, which would then be favored by selection. Thus, selection need not be the purely mechanistic process condemned by most Lamarckians, but could include a role for the active participation of the organisms themselves.

If a few selectionists tried to humanize their theory by adopting one of the Lamarckians' most appealing images of life, there were some Lamarckians who were suspicious of the whole idea. To those who were concerned about the scientific image of Lamarckism, the idea of consciouness driving evolution smacked too much of anthropomorphism and religious overenthusiasm. Many opposed the idea altogether, just as they criticized the explicitly vitalist interpretation of the theory. Kropotkin, for example, argued against the "exaggerated interference of the animal's will in the formation of new organs of which metaphysically inclined writers have lately tried to make so much."[77] The problem was that the emphasis on the will of the animal to develop a new organ raised the old specter that had always haunted Lamarckism: the misconception that thought alone was enough to stimulate growth. No Lamarckian believed this, of course, but to many the emphasis on consciousness as the driving force of evolution by means of behavioral modification came uncomfortably close to supporting the misconception. Writers such as Kropotkin certainly wanted to see life as a purposeful activity, but they hoped to defend such a view without abandoning the search for a materialist explanation of the processes involved.

It is time to illustrate some of these generalizations by looking in detail at some key themes in the defense of Lamarckism. First, there is the heredity-memory analogy, an ideal illustration of the ambiguity inherent in the system. As proposed by Butler, the analogy had an explicitly antimaterialist bias, and opponents were quick to point out that no real explanation was given by reducing the biological function of heredity to the higher-order mental process of remembering. To escape this charge, Lamarckians had an increasing tendency to insist that both heredity and memory must work in terms of an essentially similar and purely mechanistic process of information storage. Ironically, this attempt to devitalize the analogy undermined its effectiveness as an alternative to the germ plasm concept. As a pure analogy, it suited Lamarckism very well, since it explained the gradual addition of new elements to the hereditary fund of the species and fitted in nicely with the recapitulation theory. Indeed, it made recapitulation, rather than derivation from germinal units, the central theme of ontogeny. The growth of the individual was a repetition of all the remembered instructions from the past history of the species, yet it was open to additions in its final stages as new characters were acquired. Nevertheless, the temptation to see the addition of new hereditary information as the creation of new material units in the germ plasm proved too strong. As soon as Lamarckians began to adopt this approach, they lost sight of their original emphasis on the sequential nature of the growth process. Instead of seeing acquired characters as gradual additions to growth, they began to look for evidence of the sudden creation of new units in the germ plasm, thereby abandoning the uniquely Lamarckian approach to heredity based on recapitulation.

The paleontologists of the American school certainly accepted the memory analogy, and in 1893 Henry P. Orr made an attempt to turn it into a complete theory of heredity.[78] The two leading figures who attempted to revive interest in the idea in Britain were Marcus Hartog and Francis Darwin, both of whom illustrate the complex position of the analogy in the mechanist-vitalist debate. Hartog wrote an evaluation of the Hering-Butler thesis in 1896, which he had some difficulty in getting published.[79] In it he referred to the possibility of a chemical process of information storage that would replace Butler's molecular vibrations, but conceded that nothing was known of the modus operandi. He concluded with the claim that "for the moment, at least, the problem of heredity can only be elucidated by the light of mental, and not mechanical processes."[80] If this seemed to leave open the possibility of eventually finding a materialist explanation, Hartog adopted an explicitly vitalist stance when he came to write the introduction to the second edition of Butler's *Unconscious Memory* in 1910. He expressed disappointment at Driesch's lack of support for Lamarckism and chided Semon for rejecting the possibility that living organisms might be driven by psychic powers. Semon, he claimed, would never be able to complete his "Mneme" theory unless, like Driesch, "he forsakes the blind alley of mechanisticism and retraces his steps to reasonable vitalism."[81]

Francis Darwin had long been in touch with Samuel Butler and had done his best to smooth over the effects of Butler's quarrel with his father.[82] In 1908 he chose the occasion of his presidential address to the Dublin meeting of the British Association to proclaim his own faith in the inheritance of acquired characters and the memory analogy. As a botanist, his investigations of plant movements had suggested to him that habit and unconscious memory might exist even at this level. He held that the germ cells must be in a kind of telegraphic communication with the rest of the body, thus ruling out the possibility of fixed unit characters. Weismann's theory was weak precisely because it ignored the orderly process of growth illustrated by the recapitulation theory—it demanded evolution through distortions of the entire growth process rather than through smooth additions to the rhythm of development.[83] Although Francis Darwin referred to mechanistic explanations of memory storage, he also cautioned against the simpleminded use of mechanical analogies that ignored the ability of the organism to function as a coherent whole.[84]

The British writers cited two important European discussions of the memory analogy. One was by the Italian engineer Eugenio Rignano; it was first published in 1906.[85] Rignano accepted the almost complete recapitulation of phylogeny by ontogeny and held that the parallel represented a strong argument against Weismann's germinal units. He suggested a continuous epigenetic process of growth under the control of a "central zone" in the primitive nervous system. To explain the inheritance of acquired characters, Rignano invoked the analogy of the electric accumulator, postulating the formation of accumulators of nervous energy in the central zone as the means by which new characters are "remembered." The weakness of his theory lay in the lack of any mechanism for transferring such new memories from parent to offspring. Although Rignano seems to have assumed that it would work through the creation of new germinal elements, his emphasis on recapitulation ensured that he did not give too much away to the Mendelians. In a sense, the weak link in his theory was its greatest strength. He simply ignored the question of germinal structure and concentrated instead on the integrated process of growth stemming from the memory of the central zone in the embryonic nervous system. At this level, one could imagine the gradual formation of a new memory over many generations, without worring about how the accumulation of information was passed on from one individual to the next through the germ plasm.

Despite his use of a physical analogy to explain the remembering of a new character by heredity, Rignano adopted a fairly sophisticated position in the mechanist-vitalist debate. He repudiated vitalism on the grounds that it abandoned the scientific investigation of life, but at the same time insisted that the essentially rhythmical properties of complex organic systems could not be explained in terms of physics and chemistry.[86] Although the nervous system made use of physical energy, it raised it into a form with its own unique properties. In effect, Rignano adopted a position essentially similar to that later advocated by Lloyd Morgan and others under the name "emergent evolution," in which entirely new properties were supposed to be exhibited by matter when organized into complex systems.

The most influential exponent of the memory analogy in the early years of the new century was the German zoologist and anthropologist Richard Semon, whose book *Die Mneme* first appeared in 1904. Semon's policy was not to look for detailed physical explanations, but to point out the extensive range of mnemonic phenomena throughout the living world. Heredity, like any other such phenomenon, worked through the building up of "engrams," or enduring latent modifications of the irritable organic tissue. Since he did not suggest a mechanism for this process, Semon was criticized for implying that vital or even psychical forces were involved. He did claim, however, that the process must be a physiological one at bottom, insisting that his theory would be found "to be based solely upon causality, and to require no help from either vitalism or teleology."[87] He similarly repudiated Driesch's vitalism and proclaimed that the future would reveal purely physical solutions to all the problems of organic behavior.[88] In a footnote (omitted from the English translation) Semon also attacked Samuel Butler and the "psychoLamarckians" who advocated the attribution of psychic powers to living organisms.[89]

This repudiation of vitalism, however, involved a weakening of the original Lamarckian position. Semon certainly accepted the inheritance of acquired characters and discussed the latest experimental evidence at great length. Yet he seems to have had little interest in the recapitulation theory and tended to treat the creation of new engrams as a process that simply built up new fixed units in the germ plasm. He was concerned more with the remembering of the information than with its acquisition. He even claimed that it did not really matter whether the environment acted directly on the germ plasm or by provoking changes in the soma, although he certainly believed that the latter was more probable.[90] Several of the experimental illustrations he chose depended upon the sudden acquisition of characters that seemed to have no adaptive significance—exactly the kinds that were often explained as the result of a direct stimulus to the germ. In addition, Semon went to some length to show that his theory was compatible with Mendelism, implying that each new engram behaved as a fixed unit in the germ plasm.[91] The fact that Semon could give no explanation of how such units could be gradually built up meant that his examples of the inheritance of acquired characters would be seen as only minor exceptions to the normal rule of the fixed transmission of genetic units of information.

Lamarckians did advance certain observational and experimental evidence in support of their theoretical perspective. Consistent with their general philosophy, most preferred evidence suggesting a positive, adaptive response of the whole organism to changes in its environment. Such effects were not confined to use-inheritance in animals, but might include the kind of physiological response that could be found even among plants. As long as the acquired character was adaptive, evidence of its inheritance would support the Lamarckian claim that the purposeful response of life to its environment was the driving force of evolution. Unfortunately, while there was plenty of evidence that organisms did respond to their environment, it was here that evidence for the actual inheritance of the acquired characters was least satisfactory. Those experiments that were most widely accepted as showing the creation of a new hereditable character tended to involve nonadaptive changes, and were thus interpreted by Weismann and his followers as due to the direct stimulus of the environment upon the germ plasm. This was not a Lamarckian effect, strictly speaking, because the character was not first acquired by the soma. Furthermore, since the result was nonadaptive, the effect did not support the claim that evolution is a purposeful process. On the contrary, it made the organism subject to the blind caprice of the mechanical interaction between the environment and its germinal constitution.

Although most experiments tended to concentrate on animals, a popular line of indirect evidence was derived from plants. Its leading exponent was an English clergyman, George Henslow, whose support for Lamarckism was meant to preserve a role for teleology as a means of linking science and religion. Impressed in his early career by Paley's natural theology, Henslow found it impossible to see chance variation as the origin of adaptations.[92] Nor could he see how selection could be effective, since the struggle for existence eliminated most of its victims at an early age, before any adaptive characters were formed.[93] At this stage, survival was a matter of luck, not fitness. Henslow's *Origin of Floral Structures* (1888) developed the claim that the shape of flowers has been determined by the cumulative inheritance of distortions caused by the visits of insects. This was a classic example of the mistaken assumption that if a structure looks as if it has been shaped by a certain influence, then the direct action of that influence must have created it.

Henslow now began a series of investigations into what he called the "self-adaptation" of plants to their environment. His main line of evidence was the ability of many species to adapt themselves to arid conditions by acquiring a thicker external covering, frequently one with spines or bristles.[94] Individuals of many species, when grown in artificially dry conditions, acquired exactly those characters that are normal for desert species. The desert forms could not have been created by selection, since individual plants in such conditions are so widely separated that there can be no competition between them.[95] Therefore, one had to assume that the effects of self-adaptation could be accumulated by inheritance. Henslow now insisted that there are no indifferent or random variations upon which selection could act. The environment itself "induces a plant to form definite, and not indefinite variations in nature." Furthermore, "definite variations are *always* in the direction of adaptation to the environment."[96] Henslow added that Romanes had admitted to him in a letter that if he could prove this, he would destroy Darwinism. Meanwhile, a controversy developed in the pages of the popular journal *Natural Science*. Wallace criticized Henslow for simply assuming that the variations were inherited, and suggested that spines were no advantage in desert conditions.[97] Henslow's reply contained several curious points: He conceded that those individuals who survive have definite variations (which would seem to give the whole game away) and insisted that natural selection was a teleological mechanism because it required the production of adaptations in anticipation of their use.[98] In a similar controversy a few years later, Henslow complained that too many people treated natural selection as an a priori truth that could not be questioned.[99] He did, however, concede to Romanes that "self-adaptation" was not a vera causa, presumably implying that the physiological mechanisms that produced the changes would have to be investigated.[100]

The importance of physiology as the key to self-adaptation was also stressed in a survey of the evidence derived from the vegetable kingdom by Peter Kropotkin in 1910.[101] Although he did not have Henslow's theological axe to grind, Kropotkin was nevertheless convinced that living structures do have the power to respond in a purposeful way to their environment, and that this power was the key to evolution. In addition to Henslow's work, he cited experiments by the French botanist Gaston Bonnier showing that plants raised in alpine conditions acquired adaptive characters.[102] Bonnier himself did not stress the Lamarckian aspect of his work, and Kropotkin realized that the crucial question was whether or not the new characters are inherited. He argued that they must be, because the effect seemed to increase if several generations in succession were raised in the new conditions. At the same time, however, he conceded that plants taken from the mountainside and grown at a lower altitude soon lose their peculiar characteristics, thereby exposing the weakness of this whole line of argument.

As soon as one moved to the animal kingdom, this opened up the possibility that active behavior might control the organism's adaptation to its environment. Some Lamarckians referred to the work of H. S. Jennings, which demonstrated purposeful behavior even among the lowest forms of animal life—although Jennings himself preferred organic selection as the explanation of how such adaptive responses are accumulated.[103] It was among the insects that some of the most convincing work was done on the environmental stimulation of new hereditable characters, although the changes under study almost invariably consisted of fairly rapid alterations of the organisms' constitution that were of no adaptive significance. Weismann himself had set the scene in his early career with a study of the effects of higher temperatures in producing different colors in butterflies, an effect confirmed by Max Standfuss and others toward the end of the century.[104] Naturally enough, Weismann went on to insist that this was not self-adaptation, but a direct effect of the higher temperature upon the germ plasm of the developing insect.[105] The environment did play a role in eliciting the germinal variation, but not by first influencing the soma, as the Lamarckians assumed. Other writers suspicious of pure Lamarckism allowed the possibility of such an effect.[106] Since they were assumed to be nonadaptive, these examples of "determinate variation" could be regarded as small-scale illustrations of orthogenesis, not Lamarckism. Theodor Eimer, the leading proponent of orthogenesis, certainly insisted on taking this effect as illustrating a single stage in the long-range nonadaptive trends that he regarded as a major feature of evolution.[107]

Experiments with the higher animals gave equally confusing results.

Despite protestations that their theory did not require the inheritance of harmful effects such as mutilations, many Lamarckians continued to cite this kind of evidence. The classic case was still that of Brown-Séquard's work on the inheritance of epilepsy in guinea pigs subjected to lesions of the nervous system. In 1890, fifteen years after the original experiments, Romanes went to Paris in an effort to duplicate the original results. He was unsuccessful, but conceded that he had not done enough experiments to disprove Brown-Séquard's original report.[108] In his biography of Brown-Séquard, J.M.D. Olmsted attributes the original epileptic effect to a neurotoxin produced by lice, and assumes that its apparent inheritance was due to coincidence in an animal known to be vulnerable to this condition in the laboratory.[109] After 1900, these experiments began to fall into the background as a new generation of laboratory workers launched an assault upon the problem of Lamarckism. Many of these new experiments still involved a search for the inheritance of detrimental effects. Guinea pigs and other animals were subjected to a range of abnormal conditions including alcohol poisoning and rapid spinning on a turntable. Positive results for the inheritance of the resulting characters were reported in a number of cases.[110] However, the fact that the acquired characters were nonadaptive once again undermined any truly Lamarckian explanation. Even those biologists reporting the results accepted that the deleterious external influences merely damaged the germ plasm in some way. This was a long way from the original Lamarckian claim that positive new characters could be built up in the germ plasm in response to adaptations acquired by the adult body.

Harmful effects may have been favored for laboratory experiments because they were so obvious that their inheritance would be easy to detect; but by adopting this approach, the Lamarckians gave too much away to the new science of genetics. They also undermined the emotional appeal of their theory, which had depended upon the assumption that only positive effects would be incorporated into evolution. It is precisely for this reason that the experiments of Paul Kammerer became so crucial, since they appeared to show the inheritance of suddenly acquired *adaptive* characters. I shall discuss Kammerer's work in detail below, after first noting some of the indirect evidence that was offered for the inheritance of positive characters in animals. It is important to note that use-inheritance, theoretically the most important Lamarckian phenomenon, was recognized by all to be the most difficult to detect. Indirect arguments such as those offered by the paleontologists were thus all the more crucial here. An alternative source of evidence, based on the direction of hair in the coats of animals, was offered by Walter Kidd.[111]

Although Kidd claimed to have demonstrative proof that the direction of hair was shaped by the animals' movements, it is clear that he had no experimental evidence for an inherited effect. The title of a later work does, however, indicate that Kidd was aware of the implications of his claims: *Initiative in Evolution.*[112]

A good illustration of the complex interaction between theory and experiment can be seen in the work of Joseph T. Cunningham, an important but by no means typical British Lamarckian. At one point, Cunningham declared himself a disciple of Spencer,[113] although in fact his views coincided neither with Spencer's nor with the Butler schools'. He wrote bitterly against the "New Darwinism," which he saw as promoting social policies based on free-for-all competition.[114] Yet he would not side with those who portrayed Lamarckism as an instrument of social progress, since he was well aware of the extent to which natural evolution has led to degeneration. Although he accepted the traditional Lamarckian claim that new habits are the driving force of adaptive evolution,[115] his real interest was the development of nonadaptive characters both by habit and by direct stimulation of the organism's physiology. It was Cunningham who defied Wallace to convince him that any of the characters normally used to distinguish closely related species are adaptive.[116] At first a strong supporter of the recapitulation theory, Cunningham rejected Weismann's germ plasm theory largely because it did not seem adequate to explain the orderliness of growth.[117] Although he believed that recapitulation followed from the addition of adaptive steps to ontogeny, he was also tempted by the idea that regular nonadaptive trends might be built into evolution by the tendency of certain forms to vary in a predetermined direction.

As a naturalist attached to the Marine Biological Association, Cunningham's most widely quoted experiment in support of Lamarckism was performed to study the action of light on the coloration of flatfish. The detailed results were reported to the Royal Society in 1893, but Cunningham saved the theoretical conclusions for a less formal publication.[118] Exposure to light from below did not stop young flatfish from losing the color on their undersides as they adopted the adult form, but continued exposure to light did eventually produce colors on the underside of adults. Cunningham accepted the fact that the natural tendency to lose color on the underside must be hereditary, but argued that the continued possibility of an environmental effect indicated a Lamarckian origin for the character. If, however, color was lost through the removal of the light stimulus, did this not imply that the coloration of the upper surfaces was actually produced by light? This left Cunningham facing a dilemma: Many naturalists assumed that the color of flatfish had adaptive value as camouflage, but it was generally agreed that protective coloration was one of the few kinds of character that Lamarckism could not produce. He escaped by simply proclaiming that the colors are not adaptive and by postulating internally controlled variation-trends that produce regular geometric patterns in the evolution of a group.[119]

Cunningham's next major project was a comprehensive survey of secondary sexual characters aimed at demonstrating their Lamarckian origin. His theory was based on use-inheritance and the continued effect of habit in stimulating those parts of the body most used in courtship and mating.[120] He argued that the characters thus acquired are not adaptive—in effect, they are merely by-products of the mating habits peculiar to each species. Later on, he abandoned this view in favor of the belief that the hormones produced during mating stimulate the nonadaptive growth of certain parts and can affect the germ plasm to render the effect hereditary.[121] Such a theory is very different from the optimistic philosophy of Lamarckism, since it implies that evolution is directed by internal processes over which the individual organism has no control.

It was no coincidence that one of Cunningham's first moves against the germ plasm theory had been to translate a work by one of Weismann's most vitriolic opponents, Theodor Eimer. Apart from a common interest in Lamarckism, both men shared the belief that much of evolution has been governed by nonadaptive trends. The first volume of Eimer's *Entstehung der Arten* (1888), which Cunningham translated, was indeed a comprehensive survey of all the arguments in favor of Lamarckism and opposed to Weismann.[122] Eimer's position was that evolution unfolds through an interaction between the "laws of growth" built into the organism and the external stimuli to which it is exposed. He held that the positive response of the organism to its environment had led, via use-inheritance, to all the major developments in evolution. He even implied that recapitulation could be explained by the growth process somehow mimicking the external stimuli to which the organism's ancestors had been exposed.[123] He believed that animals' powers of sensation and will were derived from the irritability of protoplasm by acquisition and inheritance. Yet far from exalting the role of the individual in controlling evolution, Eimer's thoughts led in exactly the opposite direction. During the 1890s, his observations on color variation in butterflies led him to believe that nonadaptive evolution is the more important, driven by variation-trends built into the basic constitution of each major group. The theory of orthogenesis thus emerged from Eimer's original Lamarckism as he came to place more emphasis

on the lawlike regularity of growth, with the environment being reduced to a mere stimulus responsible for eliciting each predetermined
stage of development. In his social opinions Eimer also tried to play
down the role of the individual, proclaiming the necessity of each person subordinating oneself to the state. He even used the beehive as his
model for the ideal human community.[124] This suggests that the connection that has been drawn between the idealist element in Haeckel's
Lamarckism and the subsequent emergence of Nazi totalitarianism is
valid.

I shall return to Eimer's orthogenesis below, but not before com-
menting on the implications for Lamarckism of his views. The ease with
which extreme anti-Darwinians such as Cunningham and Eimer moved
from the positive image of use-inheritance toward a more pessimistic
view of internally directed evolution cannot be glossed over. The most
effective way of attacking Darwinism was to challenge the principle of
utility upon which it was based by advertising the prevalence of nonadaptive structures in nature. Lamarckism could be bent to such a
purpose because it stressed the highly organized process of individual
growth as the key to evolution. If once the emphasis was switched from
the external goal of adaptation to the internally programmed laws of
growth, a highly formalized theory of nonadaptive evolution would result. Cunningham held the internal trends down to a fairly limited scale,
but Eimer saw them as major features controlling the development of
whole phyla. As we shall see, this ultimately led to a kind of revived
idealism in which all evolution is unified by purely formal laws of development. If Eimer's views seem exaggerated, it should be remembered
that the same viewpoint lay at the heart of the characteristic philosophy of American neo-Lamarckism. Few Americans would have followed Eimer in his support for a totalitarian "beehive" society, but the
fact that such an implication could be drawn from a world view with
strong conceptual links to orthodox Lamarckism reveals the impossibility of identifying any scientific position with a particular philosphy
or ideology.

Eimer was probably the most active German naturalist in the fight
against Weismannism, but the Lamarckian position also drew support
from the cytologists opposed to the germ plasm theory. The great
cellular pathologist Rudolph Virchow insisted that the cell can only
change in response to an external stimulus. He regarded all such disrup-
tions as pathological in origin, although they could become normal in
later generations, thus providing a mechanism for the inheritance of
acquired characters.[125] Oscar Hertwig also opposed the germ plasm
notion with his theory of "biogenesis." He maintained that it was the

interaction of the cells derived from the egg that controls the development of the organism, and insisted that the "idioplasmic unity" of all the cells in the body made the isolation of the germ plasm impossible. Hertwig actively supported the inheritance of acquired characters and quoted the experimental evidence in its favor.[126] He was, however, suspicious of the belief that variation is a process of addition to growth, holding that the incorporation of a new character can only be achieved through a reorganization of the whole organism. For this reason he rejected the recapitulation theory and argued that the early structure of the embryo is a necessary step toward its final goal, a step required by the laws of growth rather than by any necessity to repeat ancestral forms.[127] Such a view cannot have pleased the orthodox Lamarckians, many of whom were committed to the recapitulation theory. Hertwig remained a leading opponent of Darwinism, though, and launched a major attack on the selection theory during the war years.[128]

The new direction given to the study of heredity by the rediscovery of Mendelism required a corresponding change of emphasis in the defense of Lamarckism. Providing experimental evidence for the inheritance of acquired characters now became all the more crucial, and the Lamarckians redoubled their efforts in this area. It was in these circumstances that a young biologist with a remarkable skill in the breeding of reptiles and amphibians came to Hans Przibram's newly founded Institute for Experimental Biology in Vienna. His name was Paul Kammerer, and thanks to Arthur Koestler's reconstruction of the case of the midwife toad, he is by far the best-known figure in that last-ditch experimental defense of Lamarckism. In the 1920s, the exposure of an apparent fraud in Kammerer's experiments with the midwife toad left the Lamarckians completely vulnerable to the charge that their theory was nothing but wishful thinking that could never be turned into hard science. Kammerer had been hailed as one of the leading experimentalists in the field, his results widely cited because they seemed at last to confirm the operations of use-inheritance—the most crucial and yet the least substantiated aspect of Lamarckism. At the same time, we should not fall into the trap of seeing him as an isolated figure desperately trying to stem a tide that had turned completely against him. Whatever the situation after the war, in the years before 1914 Kammerer was a leading contributor to a still very substantial Lamarckian movement.

Przibram had founded his Institute in 1903 as a deliberate effort to stimulate work in the new experimental biology. When Kammerer began his studies there, he was—according to his own later claim—under the influence of Weismann and Mendelism.[129] Although his own ex-

periences soon drove him toward Lamarckism, he always acknowledged the validity of the Mendelians' results and tried to show that the
two approaches were not incompatible. Indeed, the great limitation of
his writings was that they offered no theoretical alternative to Mendelism. Kammerer believed that heredity normally worked through unit
characters, and his efforts to show how such units could be built up
through Lamarckism were never very satisfactory. For all his appreciation of the optimistic implications of Lamarckism, Kammerer never
undertook the search for a workable alternative to the hereditarianism
that was becoming the basis of most biologists' thinking.

The experiments that brought Kammerer to fame were performed
with the midwife toad, *Alytes obstetricans,* and the salamander, *Salamandra maculosa.* (A third series of experiments on ascidians, or sea
squirts, was not reported before the war and never received much attention.) It was the midwife toad experiments that led to his downfall,
although Koestler quite rightly points out that in the end Kammerer
was freely conceding that these results did not prove Lamarckism.
However, this later concession should not be taken too seriously: In
the prewar years Kammerer was certainly prepared to stress the midwife toad experiments in popular accounts of his work and its implications.[130] The experiments were first described in 1906 and then more
fully in 1909.[131] Essentially, Kammerer persuaded the midwife toad
to mate under artificially humid conditions. Since the species normally
mates on dry land, the males have lost the mating pad on the forelimbs
that normal toads use to grasp the female while mating in water.
After several generations of being forced to mate in water, Kammerer's
specimens acquired mating pads, which appeared to be transmitted by
heredity. In the most optimistic interpretation, this was an illustration
of the inheritance of a structure produced directly by the organism's
own efforts: "The males, probably on account of the difficulty of
clasping the female in the water, have developed as an adaptation coarse
swellings on their thumbs."[132] As Kammerer was forced to concede,
however, since the character is normal in all other species of toads, it
could be merely an atavism, or a reversion to an ancestral character that
has been obscured by the recent evolution of this particular species. This
would account for the consistency of the "new" character, which appeared to breed true according to Mendel's laws. At first Kammerer even
tried to turn this concession to his advantage. The experiments showed
that "acquired characters not only become transmitted, but in mixing
with unchanged characters they follow Mendel's laws. The acquired
character, therefore, has a chance to come forth pure and in a certain
percentage, from the mixture of characters, and is thus preserved."[133]

In other words, Mendelism allowed the Lamarckian to concede that not all individuals would acquire the new character, without being exposed to the kind of swamping argument that Fleeming Jenkin had used against Darwin!

There were several experiments involving salamanders.[134] In the first, Kammerer forced *Salamandra maculosa* to breed under cold, dry conditions, where it eventually began to give birth to live young in the manner typical of its alpine relative, *S. atra*. The reverse effect was also obtained by breeding *S. atra* in water. Both effects were to some extent inherited by the second generation. A more widely noted experiment consisted of rearing *S. maculosa* on a yellow background, where its small yellow spots became enlarged to give broad bands of the same color. Conversely, when reared against a black background, the spots diminished in size and the animals became predominantly black. Both effects increased over several generations and thus appeared to be inherited. Breeding experiments showed that the artificially yellow variety did not breed true, although a naturally occurring yellow form did hybridize according to Mendel's laws. When ovaries from the naturally yellow variety were transplanted into normal females, the character still bred true, although when the same technique was repeated with artificially yellowed individuals, there was a strong tendency toward reversion.

There were many experiments involving acquired coloration among insects, but the effect could easily be dismissed as a direct and non-adaptive influence of the environment on the germ plasm. Kammerer's salamanders, however, possessed a chamelion like ability to adapt their own coloration to their surroundings. If the results were inherited, Kammerer had evidence for a kind of use-inheritance stemming from the animals' voluntary response to their environment. Here, if anywhere, one had evidence to support psychoLamarckism—the animals almost literally "willed" themselves a new color. Yet Kammerer would not follow up this lead. Even in his popular writing he made no effort to stress the element of voluntary control, so that his readers could easily have believed that it was a purely involuntary or physiological response to some change in the environment. This is not to say that he was unaware of the wider implications of Lamarckism. On the contrary, he was firmly convinced that the theory offered the prospect of rapid social improvement.

This wonderful new result, together with all those previously attained, opens an entirely new path for the improvement of our race, the purifying and strengthening of all humanity—a more beautiful and worthy method than that advanced by fanatic race enthusiasts, which is based upon the relentless struggle for existence, through race hatred and selection of races, which doubtless are thoroughly distasteful to many. . . . If acquired characters, impressions of the individual life, can, as a general thing, be inherited, the works and words of men undoubtedly belong to them. Thus viewed, each act, even each word, has an evolutionary bearing. . . . The active striving for definite, favorable, new qualities will in a like manner yield the power to transmit the capabilities which we have acquired, the activities which we have busily practiced, the overcoming of trials and illness—will leave somewhere their impress upon our children or our children's children. Even if ever so much weakened, even if only in disposition or tendency, not in completed form, even if completely concealed for generations, some reflection of that which we have been and what we have done must be transmitted to our descendants. [135]

What is apparent here is a shift of emphasis in the optimistic message of Lamarckism away from the old metaphysical speculations on the role of mind in nature. Kammerer even suggested that Semon's mneme theory would be better expressed by drawing an analogy with the physical property of inertia rather than the psychological function of memory. [136] The old tradition of concern for consciousness and design had now been converted into a more humanistic optimism related mainly to social problems. The message still had great appeal, but it had emancipated itself from what had been the chief concerns of the Darwinian era. In his concern to translate Lamarckism into the world of the twentieth century, Kammerer had cut the theory off from its intellectual roots.

In the absence of the old traditions, it was even more important to have a new theoretical foundation to support the experimental structure. But this is precisely what Kammerer could not provide. As an exponent of the experimental method, he had to take Mendelism seriously. Indeed, his chief criterion for judging if a character had become truly hereditary was whether or not it bred true according to Mendel's laws, and he liked to think of De Vries's mutations as suddenly acquired characters. Yet his whole approach required that new characters be gradually built up through use-inheritance. Kammerer argued that the imperfect inheritance of his artificially produced salamander colors must represent the first stage in the process by which the natural, truly heritable character had evolved. The character indeed appeared quickly, but its complete inheritance was obviously a much slower process that he could not trace in the laboratory. In his later writings, Kammerer suggested that a new character in the individual set up a kind of irritation in the germ plasm that gradually disappeared as the germ absorbed the character.[137] But this was never anything more than a loose analogy, one that evaded the central problem of how a new gene could be created from nothing. In the circumstances, it is hardly surprising that the Mendelians refused to take the results seriously. Even if they were valid, they represented merely a trivial somatic process that had no connection with the true process of heredity. A unit character either existed or it did not, and vague hypothesizing about how one might be gradually created was treated with contempt.

The new wave of Lamarckism represented by Kammerer and the other experimentalists of the early twentieth century was thus doomed to failure from the beginning. It had abandoned most of the theory's conceptual roots, and it had no new theory of heredity to offer as a replacement. Its research program did not represent a new initiative for dealing with the problems of experimental biology—rather, it was a desperate attempt to show that some vague modification of its opponents' views might be necessary. Modern scientists offer many explanations of why the belief in the inheritance of acquired characters dragged on for so long, despite its theoretical bankruptcy and its dubious experimental success. At the purely scientific level, the Lamarckians failed to understand many of the conceptual subtleties required to tackle the problems of heredity and evolution theory. They did not appreciate the modern distinction between the phenotype and the genotype, and were thus convinced that changes in the body must eventually be reflected in the germ plasm. They did not grasp the significance of treating evolution as a change in the genetic makeup of a population, and were thus prone to assume that changes demonstrated at the population level over several generations must be composed of a succession of modifications to individual germ plasms, rather than a shift in the relative frequencies of genes in successive generations. On a wider scale, the optimistic hopes for social improvement pinned onto Lamarckism explain a great deal of the wishful thinking—if not outright fraud—in the desperate efforts made to salvage the theory long after it should have been evident that none of the techniques then available were immune to criticism.

These points are all valid, of course, but they should not be advanced as though the Lamarckians ought to have been aware of their theory's deficiencies. The development of modern genetics involved significant conceptual changes that were not easily popularized. It required a considerable effort of the imagination to appreciate the inadequacies of the traditional ideas and the scope of the new ones. The geneticists themselves at first failed to appreciate the value of "population thinking"—hence the long delay in the synthesis of genetics and the selection theory. The lack of emphasis on a populational approach by geneticists must also have limited their ability to pinpoint some of the weaknesses in the experimental techniques used by those trying to demonstrate the inheritance of acquired characters. Lamarckism would only gradually become less plausible as its opponents became more sophisticated in their power to explain away the effects that could be demonstrated in the laboratory. To do this they had to invoke complex processes within the gene pool, but in the early decades of the century the theoretical subtleties required to conceive of such processes were not appreciated by anyone on either side of the debate, and Lamarckism was thus less easy to discredit.

In contrast, the distinction between the genotype and the phenotype was implicit in the very origins of Mendelism. The phenomena that had been demonstrated by Mendel himself showed that the genetic constitution of the organism (its genotype) is not necessarily reflected in its physical appearance (its phenotype). This made it easy to believe that changes occurring in the adult body have no effect on the genes, and the Mendelians thus naturally took over Weismann's dogma of the isolated germ plasm. The Lamarckians certainly failed to realize that the facts of heredity can be explained by postulating a one-way flow of information from the genes to the body of the growing organism, and that there is no necessity of a reverse flow allowing bodily changes to be reflected in the genes. But this was precisely the key point on which the new hereditarian philosophy pioneered by Weismann contradicted the traditional view of the nature of the living organism. The concept of a germ plasm totally isolated from the rest of the body seemed completely alien to the belief that the organism is an integrated, self-regulating system. Such a belief had long been accepted by most naturalists, and it remained the basic philosophy of Lamarckism. It is not necessarily true that the organism's powers of self-regulation must extend to the genes upon which it was patterned. Modern geneticists can show that such a reverse flow of information is theoretically superfluous, and they believe that this postulated flow is factually nonexistent. Nonetheless, the exemption of this one path of information flow was difficult for anyone trained in the old philosophy of nature to imagine, and it was this "failure" of the imagination that caused the search for factual confirmation. To show that there need be no such link between the adult body and the genes required far more than a few experiments with negative results: It required a complete demonstration of the old system's inability to cope with the problems that biologists now found most interesting.

In effect, such a change required a scientific revolution leading to the acceptance of a whole new conceptual system, or world view. Weismann and his followers had pioneered such a system and had insisted on it as a dogma long before it was sophisticated enough to provide itself with a defense against all empirical counter-demonstrations. The geneticists inherited the dogma and gradually improved their ability to undermine the Lamarckian arguments, but debates between rival conceptual systems cannot be resolved by experimental demonstrations. They are resolved by one side or the other's ability to convince the majority of scientists that its own views form the basis of a more fruitful research program. The geneticists were successful in converting the majority of laboratory biologists to their viewpoint during the early years of this century, but they did not achieve this success by proving that no Lamarckian effect is possible. Many loopholes were left unplugged, as the succession of Lamarckian experiments was to show. Indeed, Ted Steele's modern research confirms the implausibility of any claim that Lamarckism is demonstrably false. Nor were the geneticists able to show that the inheritance of acquired characters was a completely superfluous hypothesis, despite their increasing ability to explain complex effects without conceding this violation of their basic principles. There were many areas of biology, especially in field studies and paleontology, to which the new genetics itself at first seemed inapplicable, and these areas provided a constant fund of support for the more traditional viewpoint. The geneticists' success among the laboratory biologists stemmed from the fact that Lamarckism proved unable to supply any lead in setting up a rival research program, so that its remaining adherents in this field were reduced to searching for loopholes in the successful new system. Once this point had become obvious, it was inevitable that few younger scientists would be willing to risk their careers on so peripheral an activity. Even before the war, however, a new problem had already begun to emerge for the Lamarckians: the suggestion that there might be something actually fraudulent about Kammerer's crucial experiments.

POSTWAR LAMARCKISM

As European scientists got back to work again after the war, the general attitude became more hostile toward Lamarckism. Many neo-Darwinists and Mendelians had lost patience long before, but the efforts of Kammerer and others had for a while prevented the scientific community as a whole from dismissing the hypothesis as untestable.

Efforts to improve the experimental evidence continued after the war, but each "successful" report was greeted with a barrage of criticism explaining how the results could be accounted for without Lamarckism. The inability of the Lamarckians to supply a conceptual alternative to Mendelism now left them totally exposed to the attacks of those who saw their work as peripheral to the real study of heredity. The eventual exposure of an apparent fraud in Kammerer's midwife toad experiments was the last straw as far as most experimentalists were concerned.

Outside the ranks of the experimental biologists, however, the collapse of Lamarckism may not have been quite so rapid. Some paleontologists and field naturalists continued to offer indirect support, which only gradually faded away as the power of the modern synthesis became apparent in the 1930s. Popular writers such as Bernard Shaw still hailed the philosophical message of psychoLamarckism, but there was no major Lamarckian initiative even at this level. Social theorists had by now abandoned the idea of a link between their subject and biological evolution; Shaw could do little more than rehash Butler's instinctive distaste for Darwinism; and those who were most concerned with the philosophical issues raised by biology had now turned to other approaches such as Lloyd Morgan's emergent evolution.

In a sense, the ever-diminishing effort to provide experimental support for Lamarckism was only a remnant of the old tradition. Even this attempt was dealt a severe blow by the exposure of Kammerer's midwife toad experiments. The chief critic of Kammerer's work was the geneticist William Bateson. Early in his career, Bateson himself seems to have toyed with Lamarckism, but an unsuccessful field trip to central Asia turned him against the theory; and once he became a leading exponent of Mendelism, he adopted an implacably hostile stance. The details of his campaign against Kammerer have been described by Koestler.[138] Already in his *Problems of Genetics* (1913), Bateson had given a critical account of the experiments and had hinted that the results "will strike most readers as improbable."[139] After the war, when Kammerer was working under conditions of extreme difficulty, Bateson expanded his criticism and again hinted that something was wrong with the results. He now began to concentrate on the midwife toad experiments to the exclusion of Kammerer's other work. In 1923 Kammerer visited Britain and America in an attempt to vindicate his name and gain support for further experiments. The text of his talk to the Cambridge Natural History Society was printed in *Nature,* and in the following year his book was published in New York. In 1926, however, a test carried out by G. K. Noble of the American Museum of Natural

History revealed that the mating pad on a crucial specimen of the midwife toad had been artificially marked with india ink. Kammerer insisted that the ink must have been injected by an assistant who was anxious to prevent the real mark from fading, and was supported in this by Przibram. But six weeks later he shot himself. It is not my purpose here to comment on the oft-debated question of whether Kammerer had indeed faked his experiments, but his suicide certainly undermined what little credibility remained to the Lamarckian cause.

In the postwar years, Kammerer conceded that the midwife toad results did not give conclusive proof of the inheritance of acquired characters and tried to throw the emphasis onto his other work.[140] The mating pad might be an atavism or the result of a direct action of the environment on the germ. It had even been suggested that the salamanders' "acquired" characters were themselves a reversion to a state developed during the last ice age. Sir Arthur Keith gained some publicity by suggesting that all supposedly acquired characters were in fact atavisms, thereby checkmating the optimistic social philosophy of Lamarckism.[141] Kammerer himself had attracted wide publicity during his tours of Britain and America, with newspaper headlines proclaiming the breeding of a new race of supermen.[142] Such exaggerations certainly contributed to the creation of a hostile opinion of Kammerer within the scientific community, yet they also show that there was still a strong public interest in the philosophy of social progress that Kammerer and other Lamarckians were expounding. Although Kammerer repudiated the exaggerated press reports in his English-language book, he still devoted a considerable amount of space to the social implications of his work.

Whatever one might say about Kammerer's character, he was one of the few Lamarckians to extend the hope of future progress to all branches of the human race.[143] It has been argued that Kammerer may have been discredited through a Nazi plot, precisely because his views undermined the "scientific" foundations of racism, but this must not be taken to imply that Lamarckism stood necessarily opposed to racism. In fact, Nazi race theory grew at least in part out of hierarchical elements within Haeckel's Lamarckian view of evolution, while the American neo-Lamarckians supported a similar interpretation.[144] Whatever the limitations imposed by Kammerer's repudiation of Lamarckism's traditional heritage, his lack of interest in the recapitulation theory allowed him to escape from the hierarchical image of growth and evolution that had all too often served as the basis for the claim that "lower" races are retarded in their development. His, at least, was an all-embracing Lamarckism that allowed every race to progress if the conditions to which it is exposed can be improved.

The link between traditional Lamarckism and race theory can be seen in the writings of Kammerer's leading English-language defender, the Irish embryologist E. W. MacBride. Convinced that the races could be ranked in a hierarchy with the whites at the top, MacBride adopted an environmentalist explanation of how the racial differences were produced. He argued that long exposure to poor conditions had effectively imposed an inferior status on some races.[145] Improvements might theoretically be possible, but would take too long to be of any practical use. If this were so, of course, it would effectively dash all hopes of a rapid improvement of the white race—a point that may not have been lost on those who were still tempted by the optimistic view of Lamarckism. Despite his different view of the implications of the theory, MacBride devoted himself to the defense of Kammerer's experimental demonstration of sudden evolution. In his volume on heredity, written in 1924 for a popular science series, he gave an account of the debate up to that point and insisted that many still believed the experiments to be genuine.[146] He suggested that Mendel's own results were a little too good to be true, and criticized both the concept of the gene and the mutation theory.[147] Building on the ideas of Rignano and Cunningham, MacBride suggested that the germ at best fixed only the basic form of each structure. The details of development were under the control of hormones from the parent's body that influenced the different stages of growth. Gradual changes in these hormones would give, in effect, the "inheritance" of acquired characters.

MacBride was one of the few biologists in the 1920s to lay continued stress on the recapitulation theory as the key to evolution.[148] Kammerer referred to the idea, but it does not seem to have been an important aspect of his Lamarckism—a point that illustrates how far his mechanistic version of the theory had abandoned its traditional roots. The 1920s, in fact, saw the almost complete elimination of the recapitulation theory from biologists' thinking. In 1930, Gavin De Beer's *Embryology and Evolution* completed the process by summing up the morphological weaknesses of the theory and pointing out its incompatibility with the new genetics. What had once been one of the chief indirect arguments for Lamarckism and a foundation of the heredity-memory analogy disappeared. The embryo—as von Baer had pointed out a century earlier—does not recapitulate the evolutionary history of its race, and hence there is no reason to believe that variation consists only of additions to the existing growth pattern. This triumph for the new approach to heredity and development illustrated the theoretical weakness of Lamarckism. The intellectual background upon which the old Lamarckism had been built could not be translated into the more mechanistic world view of twentieth-century biology. If Lamarckism were to survive, it would have to find its own mechanistic theory of heredity to rival Mendelism. Efforts were certainly made in this direction, but their obvious lack of success merely emphasized the theory's inability to cope with the new situation.

MacBride's suggestion (that hormones transferred from one generation to the next independent of the germ plasm might control the growth of the embryo) was typical of the hints that were thrown out by the Lamarckians in their desperate attempt to create a place for "soft" heredity. The idea had, in fact, been suggested by Cunningham before the war, and was described at length in 1921 in his *Hormones and Heredity*. Far more than Kammerer's vague "irritation" concept, this offered the prospect of a genuine theoretical alternative to Mendelism. Unfortunately, its advocates were unable to get beyond the stage of providing indirect evidence for the effect they postulated. They might criticize the theory that the gene was the absolute determinant of heredity, but their own ideas could not link theory and experiment in the same fruitful manner. There was no lack of Lamarckian experiments in the 1920s, but they were still devoted mainly to proving that the effect existed, rather than to exploring the details of an alternative mechanism of heredity. The hormonal theory of heredity was a false start, and modern Lamarckians such as Ted Steele are exploring quite different alternative mechanisms to explain their experiments. The fate of the earlier Lamarckism shows that evidence in favor of a basic point is not enough to ensure its scientific respectability. There must also be a theoretical structure capable of guiding detailed research, and if the supporters of the idea cannot hit upon a fruitful way of expanding it in this direction they will be ignored—whatever the "proof" they can offer.

Apart from Kammerer's work, one of the most widely discussed Lamarckian experiments was that of the psychologist William MacDougall on the inheritance of learning in rats. MacDougall had supported Lamarckism in his *Body and Mind* (1911), and conceded that he had an emotional predisposition in favor of the theory.[149] In 1927, some years after he had crossed the Atlantic to Harvard, MacDougall published his account of experiments that seemed to show that rats inherited from their parents the knowledge of how to run mazes.[150] In this essay he developed the theme that inheritance of learning would allow us to assign purposeful actions a leading role in the drama of evolution.[151] Given the reputation of Lamarckism, psychology was perhaps the science least able to inspire the confidence of biologists. MacDougall could make no contribution to the debate on the mechanism of heredity, and indeed the kind of evidence he was suggesting would have been incompatible with the hormonal theory. The inevitable controversy ensued, and eventually it was shown that MacDougall's results could be explained by a selective process of improving the ability of the rats to run any maze.[152]

If MacDougall's psychological experiment was something of an anomaly, there was no lack of biologists still willing to test the question of acquired characters. As Kammerer pointed out, however, many reported under titles that made no explicit mention of Lamarckism—a sure sign that they were aware of fighting a rearguard action.[153] One exception was the proceedings of a symposium on the inheritance of acquired characters published by the American Philosophical Society in 1923, in which the effects of various detrimental factors such as alcohol inhalation were reported.[154] In 1925 J. W. Heslop Harrison suggested that melanism (dark coloring) could be induced in moths by feeding them food contaminated by lead.[155] The problem with such experiments was that they could all be interpreted as evidence for the genes being somehow damaged by the poison, an explanation that was accepted by both Kammerer and MacBride.[156] While counting as the inheritance of acquired characters in the strictest sense of the term, such an interpretation gave absolutely no support to the claim that the more positive effects of use and disuse could be inherited. In other words, the best experimental evidence supported only the least interesting side of Lamarckism, which could easily be explained by a simple modification of the gene theory allowing certain chemicals to induce mutations. A more positive experiment by Heslop Harrison involving the inherited effects of changed feeding habits in the sawfly was criticized as being due to the selection of a few variants who happened to prefer the new food.[157]

Outside the ranks of the experimental biologists, there was much confusion over the origin of variation. Many field naturalists and paleontologists continued to be deeply suspicious of the new genetics, and they were still prepared to accept a role for Lamarckism because the theory offered the most direct explanation of the phenomena they studied. The field naturalist studying geographical variation, or the paleontologist tracing the gradual specialization of an organ, were both predisposed to accept the direct action of the environment as the cause of variation. The continued popularity of such views in the 1920s has been stressed by a number of writers in the most recent survey of the origins of the modern synthesis.[158] At the same time, it is suggested that field naturalists and paleontologists were reluctant to get deeply involved with evolution theory because they were aware of the gulf separating their approach from that of the laboratory biologists. A few were bold enough to insist that large-scale evolutionary trends would require mechanisms lying outside the scope of experimental investigation, but many preferred to ignore the details of how the process might work. There seemed no reason to suppose that a combination of genetics and selection would prove better able to cope with their problems than Lamarckism; in effect, however, these workers would be willing to support any theoretical initiative that showed a way out of the impasse into which evolutionism seemed to have fallen. Lamarckism was unable to provide the basis for such an initiative because of its failure in the laboratory. The experimentalists were turning inward rather than outward, desperately searching for hard evidence and unable to come up with any broad theory of how the Lamarckian effect worked that would prove useful in other fields. They gradually lost touch with the field naturalists and paleontologists who had always provided the best indirect evidence for Lamarckism. The way was thus left clear for Lamarckism to be eclipsed by the eventual synthesis of the Darwinian and Mendelian traditions.

Among the field naturalists, one of the most explicit Lamarckians was Bernhard Rensch, who in 1929 used the occurrence of geographical variation as evidence for the direct action of the environment on the organism. He appealed to phenomena such as "Allen's law," which expresses the generalization that in a widely distributed animal species the extremities (ears, tails, and so on) tend to be smaller in the colder regions.[159] Rensch later admitted that he had been predisposed to accept use-inheritance because he had not realized that selection itself could adapt a species to a new behavior pattern (the mechanism more generally known as organic selection, or the Baldwin effect). In 1928 G. C. Robson, a zoologist with the British Museum, argued that Lamarckism could not be ruled out, and noted that Kammerer's salamander experiments had not been discredited.[160] A few years later Robson put together a massive survey of animal variation in collaboration with O. W. Richards, which devoted considerable space to Lamarckism and was critical of the selection theory's ability to cope with the issues facing naturalists.[161] If most experimental biologists were suspicious of Lamarckism, it would seem that the few who did continue to report positive results at least succeeded in keeping alive the hopes of those naturalists who had refused to accept the principles of the new genetics. Nevertheless, the insubstantial nature of the experimental support must have weighed heavily on the minds of the remaining Lamarckians, creating a sceptical attitude toward the basic problems of evolution theory. The tension created a sense of frustration that would serve as a fertile ground in which the seeds of the modern synthesis would grow with remarkable rapidity, once the advantages of this new initiative became apparent.

One sign of the confusion reigning among the naturalists was that Robson and Richards felt obliged to mention nonscientific alternatives such as Bergson's "creative evolution."[162] This raises the more general question of Lamarckism's demise as a viable philosophy of nature. There was still some support from outside science, but the theory's influence on the academic community was rapidly diminishing. George Bernard Shaw borrowed the name "creative evolution" for his own brand of psychoLamarckism, which he defended with brilliant style in the preface to his *Back to Methuselah* (1924). Shaw revived Butler's passionate emotional rejection of Darwinism, declaring that if one could not disprove selection, one must still, from the depths of one's inner conviction, tell its supporters that they are fools and liars.[163] Curiously, Shaw seems to have felt that he was part of a new wave of support for Lamarckism, but in fact his claim that it represented the spiritual salvation of the evolution movement was no longer fashionable, even outside science.

The 1920s was not a decade of absolute hostility to the antimaterialism that had once fueled the support for Lamarckism, for it was at this time that the concept of "emergent evolution" was popularized by C. Lloyd Morgan and others, while Alfred North Whitehead advanced his "philosophy of organism."[164] In proclaiming that life transcended the laws of mechanics, these philosophies were expounding a theme that had once lain at the heart of Lamarckism—yet their supporters made no reference to the inheritance of acquired characters. Only Jan Christiaan Smuts's *Holism and Evolution* (1926) made explicit use of Lamarckism in its attack on mechanistic biology.[165] Lamarckism had lost much of its original status as a philosophy of nature, perhaps because more thinkers had now accepted the point that had originally led Lloyd Morgan to suggest the mechanism of organic selection. The behavior of living organisms could shape their evolution, even without the inheritance of acquired characters, by defining the channels along which it was advantageous for selection to advance. The social sciences too had lost interest in Lamarckism as a theory of how human cultures have developed—indeed, they had rejected the entire system of biological evolution as a model for the understanding of society. Lamarckism had thus lost its hold on the academic world as well as on science. The newspaper headlines that greeted Kammerer's optimistic social message suggest that this implication was still able to generate popular interest, but without proper leadership this would gradually wither away. The rigid hereditarianism of the eugenics movement had become the most popular biological philosophy of the early twentieth century. In Russia, Lamarckism would be revived in the 1940s under Lysenko, but that is another story.[166] In the West, writers such as Koestler have kept the theory alive as a fringe movement by exploiting the strong distaste for selection first expressed by Butler and Shaw, but it is only in the last few years that Ted Steele's work has raised the possibility of a scientific revival. Whether Lamarckism may yet come to play a role in biology is not for contemporary historians of science to decide, but at least it is possible to see why it failed so dismally in its original form.

5

Anti-Darwinism in France

M Y survey of Lamarckism has already drawn attention to one
French biologist, C. E. Brown-Séquard, whose experiments
on the inheritance of epilepsy in guinea pigs were widely discussed during the early phase of the debate. Brown-Séquard
was, however, something of an anomaly: Although a prominent figure in the French academic community, he had an
American father and traveled frequently in the English-speaking
world.[1] His work thus became widely known to evolutionists, not just
because it was topical, but because deliberate efforts had been made to
circulate the results outside France. The same cannot be said for the
majority of French biologists, and to a remarkable extent the story of
French evolutionism can be told in isolation from that of the freely in-
termingling British, American, and German scientific communities. The
respective merits of Darwinism and Lamarckism were certainly debated
in late nineteenth-century France, but in a context differing so greatly
from those I have already discussed that little interaction can be observed.

Many reasons have been advanced to explain the isolation of French
evolutionism.[2] The most obvious is that even though the *Origin of
Species* was translated into French, Darwinism was not responsible for
converting the French scientific world to evolution. Elsewhere, even the
anti-Darwinians were consciously reacting against an established Darwinism, which had already imposed its own structure on the evolutionary debate. In France, conversion to evolution was a more gradual
process, and indeed a surprising number of French biologists continued
either to ignore or even to oppose the basic idea of evolution into the
early twentieth century. Those who did accept the theory tended at
first not to worry too much about the mechanism of change; even Lamarck's views received little attention until the last decade of the
century. Yet the fact that a self-conscious Lamarckism arose in France

—as elsewhere—in reaction to Weismann's neo-Darwinism suggests that the French naturalists were at least aware of the debates going on outside the country. Linguistic and cultural chauvinism has been suggested as one reason why the French did not react to Darwin in the same way as everyone else, but the isolation was clearly not absolute. Other factors besides a simple unwillingness to communicate must be invoked to explain why the structure of French evolutionism was so different that even the reaction against Weismann produced no significant interaction with foreign anti-Darwinian forces.

An important factor contributing to the isolation of French biology was the nature of the academic community. A few senior men exerted vast influence through their ability to control the promotion of younger scientists, a situation that ensured an ingrown, essentially conservative attitude. Lucien Cuénot—later a temporary Darwinist—recalled that as late as 1883 evolution was not mentioned in the biology courses given at the Sorbonne.[3] Although Alfred Giard was already teaching evolution in the provinces, he was ignored in Paris, a point that illustrates the highly centralized nature of French academic life. Giard's move to Paris in 1888 put him in a more influential position and coincided with the emergence of an open debate on the Darwinian question, although there were still many who resisted the whole idea of evolution and stuck with a purely descriptive approach.

Another factor that has been suggested is the relative obscurity of the French naturalists who emerged during the middle part of the century.[4] In contrast to the physiologists, those naturalists who first heard the message of Darwinism were mostly of rather undistinguished caliber. The great days of Cuvier and Geoffroy Saint-Hilaire were past, and the new generation was content merely to transmit the ideas it had inherited. Those ideas were opposed to evolution, of course, but even more significant is the kind of science from which they were derived. The obscurity of the French naturalists is surely correlated with their failure to move into those areas of study that were most responsible for revolutionizing mid-nineteenth-century biology. The field naturalist tradition, with its concern for ecological relationships and geographical distribution, was certainly the source of inspiration for Darwin and Wallace, and the source of the evidence that converted their followers. In contrast, the French had remained true to the old morphological tradition: They would not move out to study nature in all its complexity in the wild state. This conservatism isolated them from the real impact of Darwinism, and points to a yet deeper cultural force shaping French science. There were no real field naturalists because this approach violated the rationalist image of science as an activity that must take place in the carefully controlled environment of the laboratory or the dissecting room.

The conservative forces ranged against evolutionism in general, and Darwinism in particular, were many and various. The Cartesian philosophy was still prominent in France, and although it promoted a materialist cosmology, it stressed a rationalist approach to science that was incompatible with both the field naturalists' willingness to take the complexity of nature at its face value and the Darwinian view of development by trial and error. This was reinforced by the antievolutionary arguments derived from Cuvier, but perhaps of even greater significance were the views of laboratory biologists such as Louis Pasteur and Claude Bernard, who stressed the speculative nature of research into the past and promoted an image of the living organism as a rationally ordered physiological system. The strong element of classical philosophy in the educational program of most schools and universities must have prejudiced many nonscientists in favor of the essentialist view of species derived from Plato and Aristotle. Nor should we forget the role of the church, which took an increasingly firm stand against modern heresies toward the end of the century.[5] All of these influences together constructed a formidable barrier that Darwinism could not penetrate, ensuring that those French scientists who converted to evolutionism did so for their own reasons.

When the *Origin of Species* appeared in translation in 1862, its title was changed to link evolution with the idea of progress, and the translator, Clémence Royer, included a strongly anticlerical preface, which almost certainly prejudiced the case against the book.[6] The link between evolution and materialism ensured that the theory would be seen in the context of the controversy raging over the spontaneous generation of life, and would thus suffer as a result of Pasteur's refutation of that ancient belief. Another common charge was that Darwin had merely revived the discredited ideas of Lamarck, a view rendered plausible by the fact that the few French scientists who did accept evolution offered no systematic discussion of its mechanism. It was known that Darwin did not deny the inheritance of acquired characters, and Ernst Haeckel provided the model of a prominent "Darwinist" who also included a major role for Lamarck. The early French evolutionists may have followed the same principle and allowed themselves to be called Darwinists, but this should not be seen as indicating that they had a deep understanding of the selection mechanism.

Although there was a strong element of Lamarckism in French evolutionism, there is no evidence of any immediate attempt to revive Lamarck's name as a symbol of opposition to Darwinism. When a new

edition of the *Philosophie zoologique* appeared in 1873, it had an introduction suggesting that Lamarck's ideas had not yet been the subject of
any serious debate. There is some evidence that French evolutionists
were drawn towards the inheritance of acquired characters by an interest in Haeckel's biogenetic law (the recapitulation theory), which
was based on an idea not to be found in Lamarck's own writings. It
was only when the excesses of Weismann's neo-Darwinism began to
create an obvious need for an alternative to selection that the French
began to see Lamarck in a more favorable light. Here was a fellow
countryman who could be assigned a place as the founder of evolutionism, and whose mechanism of use-inheritance would avoid all the conceptual problems arising from natural selection. Thus it was in the
1890s that naturalists such as Edmond Perrier and Alfred Giard began
to emerge as the leaders of a French neo-Lamarckism, although the
movement never achieved the kind of unified structure enjoyed by the
American school.

The lack of unity among French Lamarckians can be judged from
the writings of Yves Delage. In his widely respected *L'hérédité*, Delage
produced an extensive survey of all the competing theories available at
the turn of the century. He was critical of Darwinism, arguing that
selection of random variations could produce no permanent effect except under the control of man.[7] The formation of species was due not
to individual but to general variation—to new characters produced in
the entire population by exposure to changed conditions. Such general
variations were adaptive, since the organism could respond in a positive
manner to an external challenge. But Delage saw through to the non
sequitur that lay at the heart of traditional Lamarckism: There was
nothing to guarantee that the germ plasm would respond in the same
way as the body of the adult organism. Delage believed that although
germinal variation was produced by a direct response to environmental
change, the resulting characters were nonadaptive and bore no resemblance to the personal adaptations of the individual organisms. Thus,
the main lines of evolution were directed toward nonadaptive goals,
while the adaptability of life was always a phenomenon of the individual's growth in each generation. In effect, Delage was suggesting that
a massive proportion of the individual's character—all its adaptive
features—were derived not from inheritance but through a purposeful
response of the growing organism to its environment. "Phylogeny creates
organs without regard to function: ontogeny takes these organs and,
as far as possible, adapts them to necessary functions."[8] Similarly radical attempts to downplay the role of heredity were made by a few
Lamarckians elsewhere, although it is difficult to see how this view

could be reconciled with the relative stability of the vast number of wild species.

A somewhat more orthodox Lamarckian was Edmond Perrier, who was appointed to a post at the Muséum d'histoire naturelle in 1876, three years before he announced his conversion to evolutionism. His 1879 lectures on "Transformism and the Physical Sciences" suggested that evolution was part of the program by which science would reduce all phenomena to chemistry and physics.[9] Significantly, Perrier played a prominent role in toasting Haeckel during the latter's visit to France in 1878. He later wrote in defense of the recapitulation theory, arguing that tachygenesis (what the American school called the acceleration of growth) explained anomalies in the parallel between ontogeny and phylogeny.[10] The addition of new characters to growth through use-inheritance gave rise to evolution and allowed for recapitulation, but subsequent modifications of the growth process might disturb the details of the parallelism. As early as 1884, Perrier wrote a book on the precursors of Darwin, in which he maintained that evolution was the only hope of explaining how the world developed, and emphasized the role of French naturalists in the growth of evolution theory.[11] Notably, however, the book contains only a relatively short chapter on Lamarck. Although Perrier was to become a leading Lamarckian and wrote a biography of Lamarck at the end of his career, it would seem that he was predisposed to accept a Lamarckian view of evolution by his more specific interest in the recapitulation theory.

Alfred Giard first expressed support for evolution in 1874, but it was not until called to a new chair of evolution studies in Paris in 1888 that he began to exert any influence. His inaugural lecture was a historical survey of the development of evolution theory, in which Haeckel's name figured prominently on the front page, and which concluded by attributing modern evolutionism to the same three figures that Haeckel had named in the subtitle of his *History of Creation:* Goethe, Darwin, and Lamarck.[12] The introduction to his course in the following year is said to mark the beginning of support for Lamarckism as an antidote to Weismann's neo-Darwinism. Giard began by stressing the biogenetic law as proof of evolution.[13] He then drew a typical distinction between the primary and secondary causes of evolution, the primary causes being those involved with the production of variation. Without denying a role for secondary factors such as selection, Giard emphasized the ability of external influences acting directly on the organism to create new species and hailed Lamarck as the originator of this view. The dangerous exagerrations of Darwin's followers had turned the phrase "struggle for life" into an empty slogan.[14] A later article specifically attacked Weismann's "ultra-Darwinism" and referred to Brown-Séquard's experiments as convincing proof that the inheritance of acquired characters did occur.[15]

Like Perrier, Giard opted for an explicitly materialistic world view derived from Descartes's mechanical philosophy;[16] but in France, materialism did not carry the same implications as it did in Britain. Darwin had been dismissed as a materialist because his trial-and-error philosophy eliminated the argument from design. Because of its link with the highly successful tradition of French physiological research, however, the materialism urged by Perrier and Giard was meant to uphold an explicitly purposeful image of how the living organism functioned. The physiologist could show that the material organization of the body was capable of responding in positive ways to external changes in the environment—in Claude Bernard's terminology, it acted to maintain the integrity of its *milieu intérieure*. Yvette Conry has, in fact, argued that Bernard's physiology played an important role in promoting French acceptance of Lamarckism.[17] This is certainly true, although it represents only a particularly obvious example of the more general fact that Lamarckians everywhere saw the organism as a self-adjusting, highly integrated system, and believed that these powers were essential to evolution.

It has been suggested that Lamarckism was able to flourish because its adherents could not appreciate the modern distinction between the phenotype and the genotype.[18] The point is valid enough, but it is important to understand that the existing traditions within biology at first made such a distinction virtually unthinkable. The concept of the organism as an integrated self-adapting mechanism, a notion to which French naturalists in particular were exposed because of Bernard's success in physiology, promoted Lamarckian thinking in two different ways. By emphasizing the organism's own powers of adaptation, it made it very difficult to believe that these powers were not involved in the long-range adaptive trends of evolution. It also seemed unlikely that within such a highly integrated system there could be included a germinal material that was totally isolated from the rest. Today it is clear that neither of these points is sound; the genes store the information of heredity, and there is a one-way flow of information from the genes as the growth process creates a self-regulating organism. Secondary factors such as selection can be invoked to explain the genetic makeup of the entire population, but not a modification of the individual's genes. The arguments supporting this position are eminently logical—however, we are not dealing with logic here, but with deep-seated beliefs that exert a far stronger control over the imagination than mere logic. The destruction of Lamarckism required a major change in biologists' thinking, and such changes cannot be accomplished quickly. Although Bernard himself had emphasized the need for all hypotheses to be checked experimentally, Lamarckism would never be disproved by negative results in the laboratory, especially when there were at least some positive cases being reported. The additional strength given to the physiological point of view by Bernard may account for the fact that a transition that took several decades elsewhere proved impossible to achieve in France.

Outside France, the belief that the integrated character of the organism guided evolution all too often led biologists towards a vitalist philosophy of life—but perhaps because of its more explicit link with physiology, French Lamarckism was generally associated with materialism. This connection is brought out most clearly in Giard's pupil, Félix Le Dantec. Originally trained in mathematics, Le Dantec wanted to reconstruct biology along strictly materialist lines. Many of his books explored the philosophical as well as the purely biological aspects of this program. He accepted Lamarckism as a necessary consequence of his "law of functional assimilation," which expressed the body's ability to respond to its environment in a purposeful way. Although Le Dantec worked within the physiological tradition, he rejected Bernard's claim that tissues are regenerated in repose, on the grounds that this would make the Lamarckian law of use and habit incomprehensible. Tissue was built up, not destroyed, while functioning, and this functional assimilation explained the positive response to environmental challenge.[19] Le Dantec always insisted that the struggle between Lamarckism and Darwinism was unnecessary, since each contained a part of the truth. He had his own unique way of trying to synthesize the two approaches, although he always believed that the Lamarckian insight was the more valuable in helping to explain the course of evolution.

Le Dantec's ideas were developed in his *Lamarckiens et Darwiniens* (1899). In it he launched a blistering attack on Weismann and the idea that heredity was under the control of representative particles, which he compared to the old preformation theory.[20] The supposed power of these particles to control the growth of the organism was a "determinative virtue" just as meaningless as the "dormative virtue" invoked by Molière's doctor to explain the powers of opium. Le Dantec proposed a biochemical theory of heredity, in which all the properties of the fertilized ovum played a role in determining the organism's development. Changes in the parents must be reflected in their eggs or sperm, thus allowing for the inheritance of acquired characters.[21] In a brief excursion to Darwinian territory, he argued that at least some cases of mimicry could be explained as the hereditarily fixed remnants of an originally voluntary imitative faculty.[22] Tucked away at the very beginning of the book, however, was a brief statement of the idea that Le Dantec was to exploit as the basis of his "reconciliation" of Lamarckism and Darwinism. Natural selection did operate, he claimed, but only on the cells of the individual organism, ensuring that only those cells best fitted to the internal environment would survive.[23] This was to become the basis for the process of individual adaptation upon which Lamarckism was supposed to rest.

Le Dantec's *Éléments de la philosophie biologique* appeared first in its English version under the title *The Nature and Origin of Life*. Here his materialist philosophy of life was developed at length, with Le Dantec calling for more research at the colloid level to help unravel the mysteries of physiology. Again he wrote against Weismann and in favor of the belief that there is an interaction between the organism and the hereditary system. The possibility of natural selection acting on the cells making up the individual organism was now developed into the chief explanation of life's ability to adapt to new circumstances. Le Dantec used the example of a sheep exposed to anthrax bacteria. If the sheep survived, it would do so through a process of natural selection among its cells—those that through natural variation were resistant to the disease would eventually come to make up the whole body. Thus, what was an acquired character for the organism—resistance to a disease—was produced through an internal process of selection. This was the closest that any Lamarckian of the time came to Ted Steele's modern concept of "somatic selection."[24] The idea was never taken up, perhaps because Le Dantec went no further than to assume that the germ cells would participate in the process and provided no mechanism that was intelligible within the context of the new genetics; but it may also be the case that the idea too obviously violated the traditional Lamarckian concept of the organism. Le Dantec himself still believed in the organism as an integrated system: Natural selection among the cells was simply the means by which the system adjusted to a new equilibrium. In this way he managed to preserve the essence of the Lamarckian viewpoint.

> Thus, by a subterfuge, we apply the Darwinian method, not directly to the living individuals, but to the smallest independent units which go to constitute such individuals; and thus we patch up an agreement between Darwin's theory and that of Lamarck—between selection of chance variations *after the fact* and direct adaptation. But it is only a subterfuge, and with it we really adopt whole and entire Lamarck's idea—that the medium acts on the inmost tissue

of a living being by means of the mechanism which that living being is.[25]

For all Le Dantec's confidence, however, it may be that his idea seemed too similar to the Darwinian philosophy of trial and error for it to gain any place in the Lamarckian tradition.

By the time Le Dantec wrote *The Nature and Origin of Life*, Lamarckism was facing a serious challenge from the rediscovery of Mendelism and the belief that evolution proceeds through sudden mutations. Le Dantec contrasted De Vries's mutations unfavorably with the "patient adaptations" of Lamarckism. The mutations that had actually been observed were anomalous cases, not typical of normal evolutionary processes. Following Giard, he suggested that there might be a series of stable equilibria through which a species might evolve, with fairly rapid transitions from one to the next.[26] Alternatively, the original characters might have been the product of a symbiotic relationship between two species, a plant (in De Vries's examples) and a microbe. If the microbe were destroyed, sudden changes would certainly appear in the plant species thus deprived of its partner. In his *La crise du transformisme* Le Dantec repeated this view and also suggested that the discontinuously varying characters studied by the Mendelians were of only ornamental value to the species. Adaptive characters would obey different laws and would be subject to Lamarckian evolution.

If Le Dantec tried to synthesize Lamarckism and Darwinism, an even more imaginative approach to the same possibility had been proposed by Elie Metchnikoff in 1892.[27] He postulated a role in evolution for phagocytes—cells within the body designed to destroy intruders such as parasites and to eliminate useless structures. This idea was subsequently picked up by Lucien Cuénot, who had originally been the only strong Darwinist in France and who saw Metchinikoff's theory in purely Darwinian terms.[28] Eventually the influence both of Delage and of William Bateson forced Cuénot to abandon his Darwinian position and, in effect, anticipate De Vries's mutation theory. He now held that selection could not create a new organ by gradually adapting the species to new conditions. Instead, the species would have to wait until random mutation threw up a form that was "preadapted" to the new situation. Of all the French naturalists of the time, Cuénot was the best prepared to accept the possibility of a random element governing the history of life. Much later in his career, however, he abandoned this position for a more typically French teleological viewpoint.[29]

The early experiments on the appearance of genetic mutations in the fruit fly, *Drosophila,* were at first given a very different interpretation in France. A. Delcourt and Emile Guyenot proposed that since T. H. Morgan and his colleagues had not ensured uniform conditions throughout their experiments, the mutations must be induced by changes in the environment acting directly on the germ plasm.[30] When his own experiments undertaken to confirm this view gave negative results, Guyenot turned at last to Mendelism. Maurice Caullery, who had adopted a Lamarckian position while studying under Giard, also turned to Mendelism. But Caullery insisted—like many geneticists outside France—that the study of heredity had nothing to do with evolution. Unlike the foreign geneticists, however, Caullery retained his belief in Lamarckism as the cause of evolution. He justified this position by arguing that evolution had now stopped—the Lamarckian mechanism had "run out of steam," leaving only the Mendelian processes operating within the existing species.[31] Exactly the same point was made by Jean Rostand.[32] By appealing to this excuse, French Lamarckism was able to survive the collapse of the movement's experimental support. An argument that was never seriously considered elsewhere served to protect the traditional viewpoint from theoretical doubts resting upon considerations that were alien to the French way of thought.

If Lamarckism survived well into the twentieth century, there was also some support for orthogenesis, or regularly directed evolution. In his *La genèse des espèces animales* (1911), Cuénot argued both for adaptive evolution by mutation and for linear orthogenesis brought about by the tendency of certain forms to mutate consistently in a particular direction.[33] He admitted ignorance about the cause of this predisposition, a confession that perhaps left room for his eventual conversion to a teleological view of evolution. A powerful force helping to create a link between orthogenesis and the vitalist philosophy of life was the influence of Henri Bergson. In his *Creative Evolution* Bergson rejected the materialist viewpoint, which had been so successful in physiology and which had helped to support French Lamarckism. He was more sympathetic to the psycho-Lamarckism popular outside France, on the grounds that it at least preserved a role for mind in evolution. Nevertheless, in the end he rejected Lamarckism altogether because it did not account for the orderliness of evolution, which he saw as a more fundamental indication of design. Bergson's *élan vital* expressed the basic upward drive in evolution, by means of which life fought against the restrictions of the matter in which it was clothed. This underlying purposeful force was aimed at no predetermined goal, but each of its positive efforts tended to follow a consistent path until it reached the limits of its powers. Bergson explicitly supported this claim by referring to the work of Theodor Eimer and Edward Drinker Cope, leading scientific supporters of orthogenesis.[34] Bergson was widely read in the early part of the century,[35] and his philosophy represented a parallel influence to that of Lamarckism, promoting a slightly different but equally anti-Darwinian view of evolution.

French biology has continued largely isolated from the evolutionary theories developed elsewhere; neither Mendelism nor even the later revival of Darwinism were able to penetrate the cultural and academic barriers. Even today, it is said that the majority of French biologists refuse to take the modern synthesis of genetics and the selection theory seriously. The two major exceptions have been P. L'Héritier and G. Tessier, whose training as mathematicians allowed them to escape the orthodoxy imposed so rigidly upon the community of academic biologists. In a sense, Darwinism was never eclipsed in France, because there was nothing to eclipse in the first place. Nor has there been any major flowering of Darwinism in the mid-twentieth century, so strong have been the barriers erected against this alien philosophy of nature.

6

The American School

WHATEVER the success enjoyed by Lamarckism in Europe, it was in America that the theory gained its most influential position within the scientific community. American naturalists did not merely adopt a neo-Lamarckism exported from across the water—they actually pioneered the movement, coined its name ("neo-Lamarckism" was Packard's term), and developed new lines of evidence to support it. In addition, there was also something characteristic about American neo-Lamarckism, so that when I use the term "American school" I am thinking of a group of scientists identified by more than just geographical location. There were Americans who were Lamarckians and yet clearly did not belong to the American school because their Lamarckism was of the more conventional European variety. Some of these figures have already been mentioned in Chapter four above. They became more numerous as Lamarckism declined and American biologists were drawn into the last-ditch experimental defense of the movement. In its heyday, however, the American school advocated a doctrine that was certainly Lamarckian in some respects, but which also included a number of characteristic ideas that were seldom taken seriously in Europe.

The identifying features of the American school must be defined with some care. It has been suggested that the movement arose in part out of a desire to retain the link between science and natural theology;[1] but while it is true that many Americans did go out of their way to develop the religious implications of Lamarckism, there were some who did not—and in any case, this was also a characteristic of Butler's approach in Europe. On the scientific side it has also been suggested that the Americans tried to emphasize the discontinuity of evolution. This is partly true, and it is certainly a rather puzzling characteristic. The normal image presented by Lamarckism is of the gradual accumulation of slight changes, and the mechanism would seem an unlikely foundation for a theory of evolution by sudden steps, or saltations. The question leads, in fact, to the truly distinguishing feature of the American school: its vision of evolution advancing step-by-step along a regular pattern of development mirrored by the embryological growth of the individual organism. The recapitulation theory was always associated with Lamarckism, but the American school grew out of a prior concern with the embryological analogy and only later adopted the inheritance of acquired characters as the explanation of how new stages are added onto growth.[2] By taking its cue from the regular, goal-directed process of individual growth, the American school retained a fascination with the regularity of development that constantly tempted them to move in the direction of orthogenesis. The *linearity* of evolution was the chief evidence offered against Darwinism by paleontologists such as Cope and Hyatt. Where the regular trends seemed to have adaptive goals, Lamarckism could be introduced to explain how habit guided the process in a consistent direction; but when Hyatt postulated linear trends driving entire groups toward nonadaptive goals and, ultimately, extinction, he was in effect founding an influential school of orthogenesis rather than Lamarckism.

The term "orthogenesis" was popularized by Theodor Eimer to describe *nonadaptive* evolution moving consistently in a single direction.[3] The changes might be stimulated by the environment, but in essence they proceeded from an internal tendency predisposing the organisms to vary in a particular direction. The American paleontologists' claim that evolution did indeed unfold in a linear fashion was often taken as evidence for orthogenesis, although strictly speaking this would only apply in the case of nonadaptive trends. Lamarckism traditionally emphasized the positive response of the organism to its environment, and a leading modern paleontologist has suggested that it was important in keeping alive an interest in the environmental determination of evolution.[4] Even in those linear *adaptive* trends studied by Cope and others, however, the role of the environment was significantly reduced. Once the organisms had adopted a new habit in order to deal with changed conditions, the habit itself would become an internal factor driving the species toward increasing specialization, whatever the subsequent fluctuations in the environment. In the extreme case, the continued effects of habit through use-inheritance might drive an organ beyond the limit of utility, and thus adaptive Lamarckism would merge into nonadaptive orthogenesis. The same would be true when unfavorable conditions began to stimulate a loss of advanced characters and a retreat back to more primitive forms, the basis of Hyatt's theory of racial senility. The Americans' concern with expounding the regularity of evolution inevitably predisposed them to look for internal directing mechanisms—physiological or behavioral—that would explain the consistency of variation.

The origins of this characteristic concern for the linearity of evolution are not hard to find: They lie in the idealist philosophy of nature imported into America by Louis Agassiz and disseminated by him from Harvard's Museum of Comparative Zoology.[5] As a student in Germany, Agassiz had absorbed the *Naturphilosophie* of Lorenz Oken, and although he toned down some of its extravagances he remained committed to an idealist view of organic relationships. All the vertebrate species were linked into a hierarchy with man at the top, a pattern that Agassiz interpreted as a divine plan unfolded through a series of supernatural creations. Its progressive, goal-directed nature was indicated by the parallel that could be seen between the history of life on the earth and the embryological growth of the highest form. Because of his Platonic image of species as distinct elements in the divine plan, Agassiz resisted evolutionism. His debates with Asa Gray and William Barton Rogers set the scene for the American reaction to Darwinism. Although Agassiz himself never gave in, there was a growing consensus that his absolute opposition to evolution was unjustified and that his own work had laid the foundations of a developmental view of nature.

Some of Agassiz's pupils, especially the anthropologist F. W. Putnam, long remained suspicious of evolution, but the majority found the lure of the new approach irresistible—as long as they could adapt it to Agassiz's idealist vision of natural order. The American school rejected Darwinism from the start precisely because it denied the orderliness of development. Their own concept of evolution started from the "law of acceleration of growth" by which new stages were successively added to individual development as a means of bringing about directed evolution. Only later did they realize that in the case of adaptive changes, the inheritance of acquired characters would explain why the stages were added on. Their Lamarckism thus emerged in response to a growing feeling that it was not enough to postulate developments whose regularity was attributable only to direct divine control. Like Butler, they internalized the Designer, making His power part of the creative ability of life to respond to environmental challenge through use-inheritance. This optimistic interpretation of Lamarckism could not, however, be applied to the nonadaptive trends, which many Americans seemed willing to accept as purely formal patterns of development. The orthogenetic element in their thought, which I shall call "orthogenetic Lamarckism," preserved an idealist vision of development that was capable of exerting its own fascination independent of traditional natural theology. Small wonder that Darwin, who had no problems with the more normal form of Lamarckism, found the regular patterns of development postulated by Cope and Hyatt unintelligible.[6]

It was the paleontologists Cope and Hyatt who remained most faithful to the orthogenetic form of Lamarckism. They studied fossil sequences bridging vast periods of time and, given the often limited number of specimens available to them, followed their natural inclination to simplify the evolutionary trends by making them appear as linear as possible. In contrast, those of Agassiz's students who turned to field studies had to account for a quite different kind of evidence. There could be no long-range trends for them; only the bewildering variety of living forms and the environments to which they were adapted. Under these circumstances it was inevitable that they would begin to lose sight of Agassiz's original viewpoint and would gravitate toward a more normal, or what might be called "environmental," form of Lamarckism. Alpheus Packard is the best example of a naturalist who began with all the intellectual trappings of the Agassiz school, but who later adopted an environmental Lamarckism that was more intelligible to Europeans. In the end, American neo-Lamarckism consisted of an uneasy alliance between two quite different approaches to the theory. As we shall see, the paleontologists remained true to the orthogenetic element; and as the inheritance of acquired characters came increasingly under fire, the pupils of Cope and Hyatt eventually turned to pure orthogenesis. Environmental Lamarckians made some effort to adapt to the new experimental approach, but as they did so their approach became indistinguishable from that of the Europeans—and no more successful.

PALEONTOLOGY AND LAMARCKISM

Three Americans are credited with discovering the law of acceleration of growth: Alpheus Hyatt, Edward Drinker Cope, and Alpheus Packard. Of the three, Hyatt and Packard studied with Agassiz, while Cope, although a pupil of Joseph Leidy, showed clear signs of Agassiz's influence. All three were involved with the journal that became identified with their school of evolutionism, the *American Naturalist*—Hyatt and Packard as founders and Cope as a later editor. It was, however, the two paleontologists, Cope and Hyatt, who made most use of the law of acceleration in their thinking and who set up the embryological analogy as the foundation of an orthogenetic view of evolution. The law did not originate as a Lamarckian mechanism. It was at first only a means of expressing how evolution could unfold as a regular addition of stages to individual growth, the existing growth process being accelerated in order to leave room for the addition. Individual variation was thus orderly rather than random: It proceeded systematically in a predetermined direction to result in linear evolution, or what would later be termed orthogenesis. As originally conceived, there was no adaptive purpose to the additions. The pattern of development had a purely formal goal, one which exactly corresponded to Agassiz's idealist view of how the history of life has been structured by its Creator.

Cope is remembered as a vertebrate paleontologist who made significant contributions to our understanding of how many groups of modern animals evolved. He is also remembered for his exploration of the fossil riches in the American West—and the resulting feud with Othniel C. Marsh.[7] It is a measure of how far Cope's ideas on the process of evolution had advanced beyond those of his idealist forebears that his genealogies were taken seriously by later scientists who did not share his Lamarckian views. He had learned at least one of the major lessons now associated with Darwinism: that evolution has no overall goal and that each branch must be treated as a separate development. Yet in his studies of individual episodes, he retained a strong element of the old idealism. For Cope, each branch of evolution had its own goal toward which all the species in that group were evolving in a series of regular stages. Originally this goal was expressed in terms of formal structures, and Cope never abandoned the belief that many aspects of evolution have no utilitarian purpose. Nevertheless, soon after he postulated the law of acceleration, he realized that there are at least some adaptive trends and that the inheritance of acquired characters could account for the addition of stages to growth in the direction of specialization for a particular way of life. His idealism thus gave way to an orthogenetic form of Lamarckism in which behavioral innovation led to linear evolution.

Cope announced his discovery of the law of acceleration in a lengthy paper, "On the Origin of Genera," in 1868.[8] In it he maintained that only the trivial characters distinguishing the various species within a genus are adaptive, and hence only they can have been formed by natural selection. The more fundamental characters defining the genus, and the series of genera that mark the course of a group's evolution, are nonadaptive and therefore must be accounted for on grounds other than selection. The similarity between the species in a genus is not a sign of common descent; instead, every species represents a distinct line of evolution passing in parallel through the same hierarchy of generic forms. Those lines that happen to have reached the same stage of development at a particular time constitute the various species of a genus. Eventually, each will add another stage to its growth and pass on to the next generic form. Thus, the hierarchy of modern genera in a group represents part of the historic plan of development through which all member species must pass (although some earlier stages may no longer be represented today since all of the parallel lines have advanced beyond them). Evolution from one genus to the next in line is a sudden process whereby all the individuals begin to exhibit the extra stage of growth at the same time. Thus, species and genera are real entities separated from one another by sharp boundaries (and by saying this, Cope preserved a key element of Agassiz's idealist definition of species). Because each new stage is added by acceleration of the existing process of individual growth, the embryology of the individual recapitulates the hierarchy of generic forms through which the species has evolved. As to *why* the pattern should be followed, Cope offered no naturalistic explanation; instead, he argued that the course of evolution was "conceived by the Creator according to a plan of His own, according to His pleasure."[9]

Cope's first paper was based on the idealist argument from design and was, in effect, a contribution to theistic evolutionism; but by the early 1870s, he had already become prepared to admit that this position was untenable. He seems to have realized that it was no longer scientifically acceptable to invoke the Creator as the sole explanation of why evolution moved in certain directions. Some naturalistic explanation was necessary, preferably in a form that did not abandon the element of orderliness so crucial for his original view of development. Cope now conceded that many important characters had been formed in response to adaptive pressures; but rather than extend the role of natural selection, he turned to the inheritance of acquired characters as an alternative utilitarian mechanism. His reasons for adopting this course were not clearly stated. Osborn's biography of Cope attributes his conversion to the influence of Herbert Spencer, by no means an implausible suggestion since the British philosopher was already known as a supporter of use-inheritance. However, the quotation given to back up this claim is derived from the introduction to Cope's 1887 book, and there is no reference to Spencer in the first detailed account of Cope's Lamarckism—his 1873 paper, "The Method of Creation of Organic Forms." In a later paper, Cope admitted that he had not read Lamarck's own works at the time of his conversion to use-inheritance.[10] He seems to have been aware of what became conventionally known as the Lamarckian mechanism, perhaps from reading Spencer, but he presented his own position as though directly derived from studies of various cases in which animals had reacted to changes in their environment. Like many of his contemporaries, Cope simply assumed that such acquired characters would be inherited, and he seized upon this as the source of a new philosophy of evolution once he had seen the need for an adaptive mechanism to serve as an alternative to natural selection.

Cope's new approach allowed him to retain his original stress on the linear nature of evolution, although in those cases in which the trends were directed toward an adaptive goal the patterns could no longer be regarded as purely formal structures. These trends were purposeful in the sense that they represented a direct response to the animals' needs as they changed their habits to overcome challenges from the environment. Instead of imposing a preordained pattern of development on each branch of evolution, the Creator had delegated to living things the power to direct their own evolution. Life could, in effect, design itself because the Creator had endowed it with a "growth force" (bathmism), which would adapt each individual's body to new habits. If such acquired characters were inherited, the changes would accumulate over many generations and the structure of the species as a whole would become specialized for the new habit. The Lamarckian mechanism worked because the acquired characters were inherited by compression back into an earlier stage of growth, thus allowing the law of acceleration to become a natural mechanism of evolution. Cope's hostility to Darwinism now increased, since having accepted an increased role for adaptation, he had to insist that selection was inadequate to explain even this aspect of evolution. The titles of his two books, *The Origin of the Fittest* (1887) and *The Primary Factors of Organic Evolution* (1896), were intended to express his belief that selection was only of secondary importance, its function restricted to eliminating some of the less successful new characters introduced by the more fundamental Lamarckian cause of variation.

Although Cope's paleontological work concentrated on the adaptive effects of use-inheritance, he also tried to explain nonadaptive evolution by postulating the direct action of the environment upon the organisms' physiology, a process he called "physiogenesis." His *Primary Factors of Organic Evolution* begins with a study of the geographical variation within species and goes on to argue that distinct varieties are formed by the inherited effects of such direct stimulation. Some Lamarckians—George Henslow, for instance—assumed that changes brought about in this way would always tend to be adaptive, but Cope realized that this need not be the case; and indeed, many experimental studies of direct action seemed to consider characters of no adaptive significance. Cope quoted extensively from a paper by Joel A. Allen, in which variation among North American birds and mammals was attributed to the direct action of differing geographical conditions.[11] Here there was no implication of linear evolution, and this was the closest Cope ever came to the more typical Lamarckian idea of variation completely dependent on the external environment. Yet he still could not completely shake off his fascination with the regularity of variation, as shown by his own study of the North American milk snake, in which all the varieties are arranged into a neat pattern expressed in a diagram.[12]

Cope's most successful exploitation of the Lamarckian principle was in his efforts to understand the evolutionary trends linking his fossil specimens. In particular he followed up a suggestion made by John A. Ryder that the shape of the teeth might be modified by the direct action of the mechanical forces operating upon them. He thus produced the first general theory of how the widely differing teeth of the various mammals have all evolved from a primitive trituberculate form.[13] For Cope, such trends were the result of "kinetogenesis"—the direct and inherited effects of motion. At the purely descriptive level this work had permanent value, since the adaptive trends he attributed to use-inheritance could also be explained in terms of modern Darwinism. Cope himself felt differently; he believed that there were certain aspects of the fossil evidence that supported the Lamarckian interpretation. Darwinism might explain the growth of horns, for instance, once they had become useful in combat, but how could it account for the early stages of growth before the horns were large enough to be of any use? Lamarckism explained the origin of such structures by supposing that the habit of butting heads stimulated the growth of hard matter at the point of contact.[14] Cope also insisted that the regularity of many trends exhibited by the fossil sequences was incompatible with Darwinism. Selection of random variations would result in an irregular process, not the steady trends he thought could be observed in the case of horns, teeth, and so on. One of his favorite examples was the supposedly regular evolution of the modern horse by specialization from its five-toed ancestors.[15] Modern paleontologists insist that the linearity of such trends was largely illusory, a product of overrapid generalization from insufficient evidence; but Cope's background predisposed him to see evolution as a regular process, and he thus proposed a Lamarckian equivalent of the formal patterns suggested in his original theory. At the same time, the influence of the idealist definition of species is also apparent in Cope's continued insistence that evolution would be a discontinuous process occurring in sudden bursts when the pressure for change built up to an "expression point."[16]

The metaphysical implications of Lamarckism that Butler expounded in Britain were developed independently by the American school. In its first issue, the *American Naturalist* proclaimed its determination to illustrate the wisdom and goodness of the Creator,[17] but it was Cope who was responsible for the most explicit formulation of the new concept of design in which God's power was internalized in nature. Originally a Quaker and later a Unitarian, Cope had strong religious beliefs, which he was not afraid to defend in public. In 1882 he developed his concept of "archaesthetism," according to which consciousness is not a product of evolution but the cause of all progressive developments.[18] It is consciousness that directs the animals' efforts toward new goals and thus controls the applications of growth energy that will shape the form of the species. In his *Theology of Evolution* (1887), the religious implications of this view were clarified. The consciousness that guides each stage of evolution was derived from an underlying mental force in the universe that Cope identified with a finite God who was seeking to express His will in the progressive advance of life toward higher levels.

Cope was not unaware of the optimistic consequences that could be derived from his theology in the field of human progress. The divine origin of the consciousness underlying evolution guarantees that the development of higher forms must take place. In man, the process has moved into a new phase in which the conscious acquisition of new characters can be guided by foresight and a knowledge of nature's own powers. The growth of material civilization symbolizes the improvement of man's mental and moral attributes through the accumulated effects of education. Such ideas were widely accepted by American social thinkers such as Lester Ward and Joseph LeConte. It was Le Conte, himself originally a geologist, who most clearly exploited the American school's philosophy of evolution to defend the claim that reformed social conditions might improve the human race. His *Evolution and its Relation to Religious Thought* (1888) expanded Cope's belief that the theological implications of Lamarckism guarantee human progress.[19] LeConte was impressed with Agassiz's contributions to evolutionary thinking and emphasized the recapitulation theory as the foundation of Lamarckism. Unlike Cope, however, he refused to deny selection a prominent role in evolution. Lamarckism was the most basic form of change and had originally been the only mechanism of evolution; but with the appearance of the higher animals, it had been supplanted by natural selection. Lamarckism has only become important once again with man, since he has eliminated selection through morality and can now make use of the residual element of Lamarckism to control his own future development.[20] LeConte was also suspicious of Cope's idea that consciousness operates at all levels of evolution, preferring to believe that true consciousness only appeared in man. Thus, although he made use of ideas derived from the American school to uphold his social policies, LeConte's biological views showed a far greater willingness to compromise with Darwinism.

Although the American school lent support to the optimistic version of Lamarckism, its origin in Agassiz's hierarchical view of development ensured that it could also be used to defend harsher social policies. The belief that some members of the human species achieve only a more primitive stage of growth could be used to dismiss any chosen group as "inferior"—including other races, criminals, and even women. [21] This is perhaps most obvious in Cope's determination to limit his optimistic predictions of future progress to the white race. No such hope was held out to the "lower" races—indeed, Cope used the technicalities of his concepts of accelerated and retarded growth to define racial differences. [22] The Negro, for instance, was in most respects closer to man's apelike ancestors. Instead of arguing that such "inferior" qualities can be overcome by exposure to improved conditions, Cope insisted that the inferior races were permanently trapped at their lower level of development. Once again the orthogenetic and idealist element of Cope's thought eclipsed the environmental side of Lamarckism. Emphasis fell not on life's ability to respond in a positive manner to external challenges, but on the orderly, hierarchical pattern underlying all evolutionary developments. The belief, learned from Agassiz, that distinct stages exist in such a hierarchy could be used to maintain the reality of species—and the permanent inferiority of some races.

The pessimistic aspects of orthogenetic Lamarckism were most fully developed by Alpheus Hyatt. Although he followed Cope into a neo-Lamarckian theory of adaptive evolution, Hyatt broke much less decisively with the idealist concept of orderly, nonutilitarian development that he had absorbed from Agassiz. For Hyatt, use-inheritance always remained subordinate to his own discovery, the theory of racial senescence, which was a classic example of the extreme orthogenetic approach based on developments that were ultimately harmful to the type. Although both Cope and Hyatt derived the law of acceleration from the embryological parallel, Hyatt emphasized senility as a definite phase in the development of the individual and hence in evolution. The belief that the pattern of evolution had a degenerative goal restricted the role of purposeful Lamarckism and threw a stronger emphasis onto the formal aspect of development. That this emphasis derived its origin from the idealist philosophy is suggested by a report that Hyatt was deeply impressed with Agassiz's exposition of Oken's *Naturphilosophie*. [23] In this way the most esoteric aspect of German idealism continued to exert an influence on late nineteenth-century biology.

The invertebrate fossils that Hyatt studied provided ideal evidence for the link between individual growth and evolution. The coiled shells of the ammonoids, described in 1866 in his first important paper, preserved their youthful forms in the interior segments of their shells. Hyatt derived the law of acceleration by noting that the new adult characters acquired in the course of the group's evolution were gradually compressed back into the inner (that is, the less mature) parts of the shell. The interior segments of geologically more recent shells preserved the youthful forms of the individuals, which could be equated with the adult forms of ancestral species. From the start, Hyatt emphasized that senility is an integral part of the growth process, and hence that the whole group ultimately degenerates into a senile phase before it becomes extinct.[24] The simple ancestral form becomes progressively more complex before it degenerates and comes to resemble the primitive form from which it began. Where Cope, however, saw degeneration only as a retardation of growth that returned the type to its ancestral character, Hyatt viewed it as a positive development due to further acceleration and as only symbolically related to the earlier form. Progress and decline were thus inescapably built into the development trend of the group, as they also were into the life processes of the individual.

Hyatt's 1866 paper made no effort to relate the progress and degeneration of the group to the demands of the environment. Primitive, advanced, and senile types were defined solely in terms of structural complexity, and there was no explanation of why the group first advanced in a certain direction and then retreated. At this point Hyatt's theory was strictly one of orthogenesis: As he later admitted, it "account[ed] for the morphological equivalence of species in different series by some invariable law of growth."[25] He did, however, note that degeneration began when the "vital powers" of the group were exhausted, and it may have been this analogy with individual growth that allowed him to develop a Lamarckian interpretation of progressive evolution. Like Cope, Hyatt soon began to concede that positive evolutionary developments did represent the acquisition of characters useful in a particular way of life. The fact that various lines of evolution within the same type moved in different directions showed that no inherent trend was involved. It was the adoption of a certain lifestyle that determined the nature of the individual's activities, and hence the course of evolution for the species. The law of acceleration of growth now described the process whereby the successive new characters acquired in response to the habits of life were incorporated into the growth process to become inherited.[26] Hyatt shared Cope's belief that the regularity of the trends exhibited by the fossil record was evidence for a Lamarckian rather than a trial-and-error mechanism of adaptation. Once a group of related forms had established a particular lifestyle, the continued and inherited effect of the same habits would drive each in parallel along the same path of specialization.

For all his work on adaptive evolution, Hyatt remained convinced that in the end a group must degenerate into senility. Eventually, the vital powers that had allowed it to flourish in response to favorable conditions would be exhausted, and the group would no longer be able to cope with the challenges posed by further changes in the environment. New characters would be produced, but now of a useless kind symbolizing the group's inability to respond in a positive way. Such senile forms could be seen in more favorable times among pathological specimens; the senile characters naturally tended to resemble those from which the group began, since the more advanced characters were sloughed off in adversity, resulting in degeneration toward simpler, more primitive forms. This element of programed degeneration inevitably created an ambiguity in Hyatt's Lamarckism. The changes were certainly produced in response to unfavorable external conditions, but if the external causation were stressed, the very generality of the effect required one to postulate that "some general physical cause acted simultaneously, or nearly so, over the whole known area of the world." More often, Hyatt tended to imply that the real cause was internal to the organisms themselves, representing a "limit to the progressive complications which may take place in any type, beyond which it can only proceed by reversing the process, and retrograding."[27] The very fact that the decline paralleled the original ascent suggested some form of internal compulsion. By the standard's of Eimer's definition, this was certainly orthogenesis rather than Lamarckism.

Hyatt seems to have believed that all evolutionary trends must lead to senility and extinction, so it is hardly surprising that one can find no support for the optimistic philosophy of Lamarckism in his writings. He rejected Cope's idea that consciousness is the guiding force of use-inheritance, apparently regarding the attribution of such a faculty to invertebrates as mere anthropomorphism.[28] If there were any transcendental purpose underlying nature, for Hyatt its efforts would indicate a purely formal concern for the orderliness of growth, not divine wisdom and benevolence. Nor could he accept the hope of inevitable social progress through Lamarckism. On a number of occasions he stressed that man himself has a number of senile or retrogressive features. The only social implication he derived from this was an argument against the emancipation of women, based on the claim that identity between the sexes was a senile characteristic.[29] Any reform that would tend to encourage similar behavior in women and men would only accelerate the decline of the human race. Once again, the idealist notion of an inevitable pattern of development that the American school inherited from Agassiz negated the optimistic aspect of Lamarckism.

Cope and Hyatt were both aware of the challenge posed by the rise of Weismann's germ plasm theory in the 1890s. They opposed all particulate theories of heredity and lent support to the analogy between heredity and memory,[30] but neither made any detailed effort to elaborate this idea into a comprehensive alternative to Weismannism. They were no doubt aware of the growing demand for experimental evidence of use-inheritance, but as paleontologists it was not their job to provide such evidence or to test their ideas on heredity in the laboratory. They belonged to the generation of naturalists that preceded the rise of experimental biology, and neither made any effort to adapt to the new movement. Cope died in 1897 and Hyatt in 1902, so they were spared the worst experiences of the degrading scramble for experimental evidence into which twentieth-century Lamarckism degenerated. The next generation of paleontologists was not so fortunate: They could not ignore the new biology, and yet their field of study prevented them from rising to the challenge. They were willing to invest only so much effort in the defense of the Lamarckian principle. When this was not successful, they retreated into their paleontological studies to explore the one characteristic which to them seemed inescapably confirmed by the fossil record: the linearity of evolution. The experimental challenge to Lamarckism thus broke up the movement by driving the paleontologists into an increasingly isolated position, from which they advanced fossil evidence for purely orthogenetic trends that were outside the scope of laboratory investigation.

This process can be seen in the work of Cope's two most important disciples, William Berryman Scott and Henry Fairfield Osborn.[31] In 1891, Scott published a paper examining the theoretical implications of his attempts to reconstruct the evolutionary history of certain mammalian groups. He referred in particular to the many cases of parallel development to be seen in the fossil record, which would lend support to Cope's belief that evolution consisted of a series of lines advancing through the same pattern of growth.[32] Perhaps the mammals had a tendency to vary in certain predetermined directions as a result of unknown "dynamical agencies" acting in a uniform manner. This would imply orthogenesis, but at this point Scott also felt that a Lamarckian element was involved. He attacked the germ plasm theory and insisted that use-inheritance must still be considered as a possibility where the fossil record supported it. The record suggested that the addition of parts "acts just *as if* the direct action of the environment and the habits of the animal were the efficient cause of change, and any explanation which excludes the direct action of such agencies is confronted by the difficulty of an immense number of the most striking coincidences."[33] Scott was clearly aware of the unsatisfactory nature of indirect arguments, but despite his refusal to link himself with any school of evolutionism, he was still supporting Cope's position.

Within a few years, however, Scott had become even more aware of the difficulties facing Lamarckism. In 1894 he wrote an article lending support to the concept of mutation as defined by the German paleontologist Wilhelm Waagen.[34] Here the term referred not to discontinuous changes (as in the more popular definition later introduced by Hugo De Vries) but to small variations occuring consistently in the same direction. Scott no longer made any reference to Lamarckism but merely insisted that some unknown force drove the evolution of each group in a certain direction. His later works continued to ignore Lamarckism, although they still lent support to the idea that evolution proceeded in parallel lines along a predetermined path.[35] In modern terminology, Scott had abandoned Lamarckism for orthogenesis.

Henry Fairfield Osborn made a more thorough—though equally abortive—effort to find a mechanism accounting for linear evolution. His earliest work explicitly followed Cope's orthogenetic Lamarckism,[36] but by 1894 he was already beginning to express doubt. He felt that the debate between Weismann and Spencer on the inheritance of acquired characters had helped to clarify the issues, but that as yet science was only beginning to understand the problem. In particular he attributed the existence of so many contradictory ideas to the "unnatural divorce of the different branches of biology, to our extreme modern specialization, to our lack of eclecticism in biology."[37] Urging that scientists consider the paleontological evidence for directed evolution, he declared that either there must be an unknown factor in heredity or—if Lamarckism were not valid—an unknown factor in evolution. This suggests that Osborn's real concern was to protect the element of orthogenesis in his interpretations, and that he was already prepared to accept that Lamarckism might not be the only way of doing this.

Osborn's first alternative was the mechanism of "organic selection," later known as the "Baldwin effect" after its co-discoverer, the psychologist James Mark Baldwin.[38] This presented habit as the guiding force of evolution—not via Lamarckism, but through its ability to define a trend in bodily structure that would be followed up by natural selection. Osborn seems to have thought of organic selection as a compromise between Darwinism and Lamarckism, but Baldwin pointed out that it really supported Darwinism since it still required only random variation. He specifically criticized Osborn for implying that variation was nonrandom.[39] Clearly this was the one point that Osborn *did* want to defend, since he immediately went on to declare that organic selection was inadequate to explain the linear evolution of characters such as the teeth, which—he claimed—could not be affected by the individual's habits.[40] Whether or not this was an adequate response to Baldwin's point, it represented a clear break with Cope's entire approach to use-inheritance. Habit was not enough to guarantee the linearity of evolution, and from this point on Osborn would not consider it a major factor, except in the case of human evolution. Significantly, he refused to accept the Lamarckians' claim that their theory would allow for a rapid improvement of the human race. Indeed, he became disturbed about the degeneration of races that might be brought about through interbreeding and allied himself with the eugenics movement in its call for immigration restriction.[41]

Osborn's later paleontological work represents one of the most consistent attempts by a twentieth-century biologist to develop a theory of orthogenesis, and it will be dealt with under that topic below. At first he persevered in his efforts to synthesize all areas of biological investigation; he studied the latest achievements of the experimental approach and suggested mechanisms that would explain how a linear variation-trend could be built into the germ plasm.[42] In the end, however, he conceded defeat and simply maintained that paleontology had evidence for evolutionary patterns that could not be detected in the short time span of the laboratory experiment. It was the morphological relationships linking his fossil specimens into orderly sequences that were his real concern. Osborn was the true heir to a tradition that strongly encouraged the naturalist to search for regular patterns of development, in the belief that only such evidence could disprove the nightmare of Darwinism. His willingness to consider a natural explanation for the variation-trends shows that he was not an idealist in the true sense of the word. Yet his preference for linear evolution must be counted as the last vestige of Agassiz's influence transmitted through the orthogenetic Lamarckism of Cope and Hyatt.

The rise of experimental biology thus drove a wedge between the orthogenetic and the purely Lamarckian interpretations of the fossil record. As integrated into Cope's theory, Lamarckism used habit as the unifying force that made the reaction of the organism to its environment a source of orthogenetic evolution. The problem was that the new experimentalism required evidence of an inherited reaction to the environment that could be demonstrated under laboratory conditions, and by the end of the century it was already becoming clear that this would be hard to find. For this reason, both Scott and Osborn turned away from the inheritance of acquired characters as an explanation of the fossil trends that were their real concern, and thus Cope's synthesis of orthogenetic and environmental Lamarckism was destroyed. As the true Lamarckians became embroiled in an increasingly strenuous attempt to find experimental evidence, the paleontologists turned back to a concern for purely morphological relationships. They remained convinced that the linear trends were genuine, but since it was now impossible to invoke Lamarckism as an explanation, they postulated orthogenetic mechanisms that were beyond the scope of laboratory investigation.

ENVIRONMENTAL LAMARCKISM

Since orthogenetic trends could only be demonstrated over long periods, this approach to Lamarckism was confined to the paleontologists. Those of Agassiz's followers who elected to specialize in other disciplines inevitably found their interests directed along new lines. They still came to see Lamarckism as the only way of incorporating adaptive evolution into an acceptable philosophy of nature, but the purposefulness of life would have to be expressed through its active power of responding to the environment and not in any abstract pattern of development. The kind of evidence studied by the field naturalist or laboratory biologist simply did not lend itself to the demonstration of linear trends. The concentration on individual cases in which characters were acquired in response to environmental change focused attention directly on the interaction itself. Each example was of a different kind, depending on the conditions, and there was little scope for the idealist philosophy of unification.

Such studies were not incompatible with orthogenetic Lamarckism, since they did not contradict the claim that a habit, if applied consistently over a vast period of time, might generate a linear trend. As Cope was able to show, the two approaches could be combined into a dual-purpose Lamarckism: The direct effect of varying conditions produced small-scale divergent evolution, while the long-term effects of habit exploited a similar effect more consistently and brought about linear trends. There was, however, one crucial difference: It was clear that the basic assumption of environmental Lamarckism could be tested, at least in principle. Unlike the paleontologist, the environmental Lamarckian could not retreat into the vague claim that the effect was too long range for laboratory confirmation. The growing emphasis on the experimental study of heredity thus hit the environmentalists particularly hard. As in Europe, they became embroiled in an increasingly urgent attempt to provide a theoretical and experimental alternative to Weismann's position. The lack of any clear success hastened the paleontologists' retreat into pure orthogenesis, while the loss of the orthogenetic support left American neo-Lamarckism with no distinguishing feature and marked the end of the American school.

Of those Agassiz-trained naturalists who adopted the environmentalist position, the most prominent was Alpheus S. Packard, Jr. He was the only member of the school who appears to have had a prior interest in Lamarck's own writings.[43] For this reason one might have assumed that he was predisposed to see the direct action of the environment as the key to evolution. In his later biography of Lamarck, Packard claimed that he was first drawn to Lamarckism through his study of *Limulus*, the horseshoe crab, as early as 1870.[44] Yet his article on the subject merely supports the law of acceleration, on the grounds that it would account for the apparent rapidity of certain phases in arthropod evolution. A slightly later article on the same topic does refer to the action of external causes to explain the extended development of *Limulus* within the egg, but still ascribes the course of development to a "deeply seated law of growth."[45] A similar case for the acceleration and retardation of growth was made in his first study of cave fauna in 1872.[46] Packard certainly did not get this idea from Lamarck, since he later admitted that the French naturalist did not know of the biogenetic law linking embryology and evolution.[47] This suggests that, like Cope and Hyatt, Packard was attracted first to the concept of evolution by addition to growth, and only later saw the inheritance of acquired characters as an explanation of the effect.

Whatever the origins of Packard's Lamarckism, he retained an interest in embryology throughout his career. In 1894 he published a lengthy argument for the inheritance of acquired characters based on the stages of growth in insects with complete metamorphosis.[48] His chief point was that each stage of growth is adapted to a different lifestyle, thus indicating the direct action of the environment. Such evidence would inevitably have turned Packard away from an interest in the orthogenetic aspects of growth and would help to undermine the recapitulation theory. Although he did postulate general Lamarckian effects influencing a wide range of species, these were always linked to the direct stimulus of the environment rather than to the consistent effects of use. Thus, he accounted for the formation of spines in cater-pillars as due to the flow of body fluids to the skin as a result of taking up a more exposed lifestyle.[49]

Although Packard is now remembered mainly as an entomologist, he was in fact a field naturalist with a wide range of interests. He under-took a lengthy study of the blind creatures that inhabit deep caves, beginning with the 1872 work cited above and culminating with a sub-stantial monograph in 1888.[50] In his later writings, the loss of the eyes was, of course, attributed to the inherited effects of disuse. Packard, like most other Lamarckians, believed that natural selection could not account for so complete an elimination of an unwanted organ—a point eventually conceded even by Weismann. Packard strengthened the case by arguing that the loss seemed to have been a comparatively rapid one. It is instructive to compare this application of environmental La-marckism with Cope's orthogenetic approach. Although both explored the continued effects of use and disuse, Cope saw habit as effecting consistent long-term change in an entire group, while Packard saw a small, isolated population of unrelated species undergoing a very rapid degeneration. It is difficult not to believe that the kind of evidence available to the paleontologist and the field naturalist, respectively, was essential in determining how each would exploit the basic Lamarckian principle.

Because of his early interest, Packard was more aware than his fel-low members of the American school that their acceptance of the inheritance of acquired characters constituted a partial revival of La-marck's views. In 1885 he used, apparently for the first time, the term "neo-Lamarckism" to denote a school of thought opposed to Darwin-ism.[51] His biography of Lamarck gave a detailed exposition of the French writer's views and a comprehensive survey of the neo-Lamarckian scene. Yet Packard did not go out of his way to stress the wider impli-cations of Lamarckism. He agreed that it rendered evolution more com-patible with the argument from design, and he shared the common misconception that the progress of civilization implied the inheritance of acquired characters;[52] but in general, he preferred to stress the scien-tific arguments, and it is possible that his caution reflected at least in part the nature of the evidence available to him. Many of his entomo-logical examples were attributed to the direct action of the environment rather than to use-inheritance (Cope's physiogenesis rather than kineto-genesis). These effects were not always adaptive, and certainly gave less support to the claim that the organism's own efforts directed evolution. Even his chief example of use-inheritance—the blind cave animals— emphasized the degenerative effects of disuse rather than any positive development. Such examples certainly helped to obscure the image of Lamarckism as a necessarily progressive mechanism of evolution.

It may also be significant that Packard refused to adopt a totally critical attitude toward Darwinism. He agreed that selection was only a secondary mechanism acting upon variations produced by Lamarckism, but unlike many of his fellow Americans he continued to insist that it did play an important role in evolution.[53] Inevitably, he criticized Weismann's germ plasm theory, although he also conceded that selection had become more important in the later phases of evolution.[54] This willingness to make an accommodation with Darwinism may again reflect his area of specialization. As a field naturalist, Packard was in a far better position than any paleontologist to appreciate the advantages of the Darwinian approach. He worked with the same kind of evidence as Darwin himself, but was convinced that it implied a more rapid process of change than selection would allow. Packard was keenly aware of the role played by geographical isolation in divergent evolution, just as Darwin had been. Indeed, he seems to have regarded isolation as a neo-Lamarckian mechanism.[55] This indicates the extent to which an environmental Lamarckian could exploit certain aspects of the Darwinian approach, although it also says something about the failure of Darwin's supporters to follow up this side of their theory in the late nineteenth century.

Packard was by no means the only student of Agassiz whose field work led toward environmental Lamarckism. Nathaniel Shaler wrote an early article on lateral symmetry in Brachiopoda from an overtly idealist viewpoint; but as he began to move into the fields of geology and geography, Shaler began to popularize a purely environmental form of Lamarckism.[56] I have already noted the contributions of Joel A. Allen, quoted at length in Cope's *Primary Factors*.[57] Allen's studies of geographical variation among North American birds and mammals illustrate the weakness of the evidence used to support environmental Lamarckism. Variations in color or size that were correlated with climatic changes were automatically assumed to result not only from a direct effect of the conditions upon the organism, but also from a cumulative effect due to inheritance. "Allen's law" described the tendency of animals at the northern extreme of a species' range to have extremities (ears and tails) of smaller than average dimensions. Such correlations were widely seen as evidence for Lamarckism, although the lack of any experimental confirmation branded the link as purely circumstantial. Many of Allen's examples were of physiogenesis rather than kinetogenesis. Since the animal's color, for instance, is not under its conscious control, any variation produced by the environment must be the result of direct action—for example, bleaching by the sun. Allen was suspicious of the claim that the effects were of adaptive value, since this might allow a role for selection. Here again we see how even the environmental version of Lamarckism could be turned into a position that would hardly support the optimistic claims of some supporters.

If the field naturalists of the Agassiz school provided the chief line of evidence for environmental Lamarckism, there was one outsider who made a unique contribution. This was the embryologist John A. Ryder, who made perhaps the most active effort to provide Lamarckism with a theory of heredity. Ryder's original interest was in the effect of mechanical forces upon the growth process of the individual, from which he became convinced that such effects are inherited. It was he who suggested to Cope that the evolution of the teeth might be explained in terms of the forces acting upon them, an idea that was to form the basis of Cope's most enduring paleontological work.[58] Ryder's own studies provided yet another line of indirect evidence for Lamarckism, although at the end of his career he began to report rather clumsy experimental demonstrations. His position as an embryologist allowed him to take a more active part than most other members of the American school in providing a theoretical alternative to Weismann's germ plasm.

Ryder does not appear to have been influenced by Agassiz's idealism. Although he sometimes referred to nonadaptive trends in evolution, he seems to have imagined that these would be imposed by purely mechanical constraints.[59] He shared Cope's belief that use-inheritance would allow the consciously directed activity of the organisms to shape the course of evolution, although this point was not developed at length.[60] Apart from his work on teeth, Ryder also studied armored creatures such as turtles and armadillos and argued that horny growths on the skin could be stimulated by contact.[61] He also asserted that the tails and fins of fishes were shaped by the way in which their elements had been fractured by movement.[62] His last paper on this subject reported experiments on the inheritance of injuries inflicted during the early stages of growth, and he claimed that a double-tailed race of goldfish had been produced by this means.[63]

As an embryologist, Ryder directly confronted the challenge of Weismann's germ plasm theory. He believed that the concept of an isolated germ plasm was contrary to the basic principles of physiology and physics. The principle of the conservation of energy ensured that all parts of the body were integrated into the same metabolism, and hence that the material of heredity could not remain isolated from changes in the soma.[64] In effect, Weismann violated the whole trend of experimental biochemistry, which was to evaluate the body's workings in the light of the physical sciences. The germ plasm was merely a revival of the old preformation theory, in which mysteriously hidden units exerted an absolute control over the organism's growth. Ryder insisted that Lamarckism was a plausible assumption that could be taken for granted unless the Darwinians could positively disprove it.

Whatever information is transmitted by heredity, Ryder was convinced that the forces acting during growth are critical in determining the final outcome. Thus, under different circumstances a new form of adult organism would result. Ryder saw his support for the role of "nurture" as a means of preserving Lamarckism, although when pushed to its extreme his approach had the effect not so much of allowing the inheritance of acquired characters as destroying the role of heredity altogether. He stated that there must be a general tendency—inherited from the parents—for the organism to grow, but that the actual course of growth may be determined solely by the forces acting on the individual in each generation. Ryder certainly claimed that this was the case for the tails of fish.[65] As he himself admitted, this made the question of the inheritance of acquired characters superfluous, since in effect the whole body is an acquired character in each generation. Ryder's theory was the ultimate extension of the Lamarckian idea that external forces acting on the growth process of the individual can determine the evolution of the species. In a sense, it was the reductio ad absurdum of this idea, and as such it proved quite incapable of stemming the rising tide of experimentally based theories of rigid heredity.

The success of these new theories in the twentieth century soon forced Lamarckism onto the defensive. As Ryder himself suspected, indirect arguments were not enough: If Lamarckism were to survive, it would have to develop a theoretical model of heredity that could serve as the basis for experimental research. Ryder was not the only American trying to develop an alternative theory,[66] but none were successful in promoting a research program within the new experimental mode. Instead, the American biologists of the next generation who preserved an interest in Lamarckism fell into exactly the same trap as their European counterparts. They tended to accept Mendelism and project their experiments solely as evidence that there might be a few exceptions to the general rule of genetic determinism. Thus, although Lamarckism was not at first excluded from the new biology, it gradually lost ground to the hereditarian theories. Experiments were conducted and successes reported, but no Lamarckian theory of heredity emerged to challenge the triumphant new genetics. The fact that Lamarckian results were described in Mendelian language ensured that they would receive little attention, even when they were not immediately rejected because of faulty techniques.

The extent of the experimental support for Lamarckism should not be underestimated. The *American Naturalist* reported a series of experiments by Charles R. Stockard on the inherited effects of alcohol fumes on rats. Stockard, along with a number of other experimental biologists, also contributed to a symposium on the inheritance of acquired characters whose results were published by the American Philosophical Society in 1923.[67] This kind of work, however, related to the least interesting aspect of Lamarckism and posed only a minor challenge to the new genetics. Stockard actually described his results as being due to the damaging of germ cells by alcohol, an interpretation that differs little from the more conventional idea of chemically induced mutations. Kammerer's visit to the United States in 1923 may also have helped to keep up interest in the experimental evidence for Lamarckism, but he too was incapable of supplying an alternative theory of heredity. By now, in fact, the American and European versions of Lamarckism had become indistinguishable, and the distinctive character of the old American school had evaporated. These last efforts to preserve a place for environmental Lamarckism reveal only that in the attempt to adapt to the new biology, the intellectual roots of the American school had been severed. The true heirs of the school were not the experimentalists doggedly trying to follow Kammerer's example, but the paleontologists who abandoned Lamarckism for an isolated position from which they continued to expound an experimentally unverifiable form of orthogenesis.

Outside the field of scientific biology, Lamarckism may have retained some support because of its optimistic social implications. In the 1920s, Kammerer's talk of improving the human race was still capable of inspiring extravagant headlines in American newspapers; but the social scientists themselves had now turned against the theory. At one time, writers such as Ward and LeConte had been the spearhead of a widespread movement calling for social reform as a means of uplifting mankind, but the new social scientists—particularly the cultural anthropologists following the lead of Franz Boas—wanted nothing to do with the legacy of nineteenth-century evolutionism.[68] They rejected Lamarckism as they rejected any implication that biological theories should guide the study of cultural development. Lamarckism was not really an issue in the great debate over nature and nurture, because neither side was interested in it any more. The hereditarians and the supporters of eugenics were opposed to it on principle, while those who insisted that human behavior is not biologically determined wanted nothing to do with a theory that still carried the implication that cultural developments could be turned into biological instincts.

7

Orthogenesis

 ODAY, Lamarckism is by far the best-known alternative to Darwinism, yet among late nineteenth-century theories it was by no means the hypothesis most fundamentally opposed to the Darwinian philosophy of evolution. Less widely discussed outside the world of science, but far more hostile to Darwinism, was the theory of orthogenesis. Lamarckism was at least a mechanism of adaptation, and it could be reconciled with the belief that evolution is an irregular, constantly branching process. In contrast, orthogenesis assumed the existence of trends that were both regular and nonadaptive, that gave rise to immense patterns of linear evolution followed in parallel by groups of related forms, and that led ultimately to extinction through racial old age. Both Lamarckism and orthogenesis denied that variation was random, but in addition orthogenesis repudiated the utilitarian claim that adaptation was the driving force of evolution. Although Darwinism and Lamarckism subordinated form to function, orthogenesis expressed the belief that purely formal trends directed evolution without reference to the demands of function or the environment. And yet there was a link between Lamarckism and orthogenesis, since both were based on the assumption that variation consisted of an addition to the growth process of the individual. For this reason, both were linked with acceptance of the recapitulation theory; but where Lamarckism postulated external influences shaping the later phases of growth, orthogenesis held that the laws of growth themselves predetermined the additional stages, creating an orderly trend in variation that did not reflect environmental influences.

Orthogenesis means, literally, linear evolution (from the Greek ὀρθός, straight, and γένεσις, generation); however, Theodor Eimer, its major popularizer, made it quite clear that he intended the term to refer solely to *nonadaptive* linear evolution. The point of this distinction was that—as the American school had shown—Lamarckism could be seen as a mechanism capable of generating linear adaptive trends. The consistent response of the body to activities demanded by a new habit would produce a gradual specialization of structure in a particular direction. It was generally argued that Darwinian natural selection could not be expected to produce such linear trends because it had to feed upon random variation, which would introduce distortions into the process. Not all scientists, however, agreed that this was a valid inference: Ludwig Plate, for instance, pointed out that once a species had begun to specialize for a particular way of life, selection could also be expected to produce a consistent trend in this advantageous direction. Plate coined the term "orthoselection" to denote such a process.[1]

Despite Plate's argument, a generation of paleontologists around 1900 grew up with the belief that the fossil evidence for linear trends was incompatible with Darwinism. In a sense they were quite right, since the American school's vision of whole series of parallel lines moving through the same pattern of development required a degree of linearity that could not be explained by orthoselection. By the end of the century, however, the inheritance of acquired characters was coming under attack from the laboratory biologists, and something more than use and habit would be needed to explain the fossil trends. Many paleontologists seized upon the fact that some of the trends seemed to end up producing nonadaptive characters as a means of arguing that their linearity was not imposed by the organisms' positive response to the environment. Instead, there must be an internal force controlling variation, one solely dependent upon the nature of the organism itself. The willingness to admit an important role for nonadpative trends thus allowed a shift of emphasis away from habit and behavior toward purely biological mechanisms of directing variation that were beyond the organism's voluntary control. The American school had, in fact, begun with the idea that the laws of individual growth required nonadaptive trends in evolution, and now the next generation was retreating from Lamarckism back to this earlier position.

Eimer himself had been a prominent Lamarckian, but his own studies of butterfly coloration convinced him quite independently that there were trends in evolution that had no adaptive purpose. He suggested that in these cases the environment might be responsible for stimulating the nonadaptive variation, but implied that the consistency of the changes demanded an internal predisposition of the organism to vary in a single direction. Several Lamarckians had cited experimental evidence for nonadaptive changes produced, for instance, by exposure to high temperatures, but their studies suggested only a unique interaction between a particular organism and a particular kind of stimulus.

Eimer's theory went beyond this by postulating that the very nature of the organism must somehow predispose it to vary only in a certain direction, whatever the environmental stimulus.

The paleontologists were in a far better position than Eimer to demonstrate the existence of long-range nonadaptive trends, and in certain respects they took the idea even further than he did. Eimer had postulated trends that were merely indifferent to the demands of the environment; but some of the strongest fossil evidence for orthogenesis was based on trends that seemed to be actively harmful, producing bizarre characters that were assumed to have caused the species' extinction. Hyatt's theory of racial senility in the ammonoids was an early version of this approach. The vertebrate paleontologists of the next generation thought that overdevelopment of the horns and teeth had caused the extinction of forms such as the Irish elk and the saber-toothed tiger. In these cases, orthogenesis took on a rather more sinister aspect: It was no longer simply an addition to the adaptive function of evolution, but something in actual competition with, or capable of perverting, that more positive function. Yet in a paradoxical sense, this later interpretation of racial senility did reintroduce the question of environmental control. Most cases of overdevelopment occurred in structures that had originally evolved for a purpose. This raised the question of the degree of freedom from environmental restraint enjoyed by a variation-trend; at the same time, it created the problem of explaining how a trend that had begun for some adaptive purpose could later on get out of step with that purpose. Such issues added further complications to those already facing the early twentieth-century naturalists who tried to reconcile the general idea of orthogenesis with the new experimental studies of heredity and variation.

Obviously, orthogenesis had little to do with the once popular idea of an inevitably progressive trend in nature. There were many different orthogenetic trends, and most of them ended in degeneration and extinction. Perspectives had changed so much that perhaps the only link with the old idea of a progressive scale of development was through the recapitulation theory. Haeckel's essentially progressionist interpretation of evolution had been based on the analogy between the stages of life's development and the growth of the embryo. For him, the sequence of vertebrate classes emerging in the course of evolution was symbolized in the goal-directed process of embryological growth in man. In contrast, the supporters of orthogenesis did not believe in an overall hierarchy of living forms that could serve as a measure of absolute progress. In this sense alone they were true Darwinians, believing that evolution consisted of many separate branches, each of which could only be understood in its own terms. Nevertheless, within each branch one could see a regular pattern of development, which many supporters of orthogenesis believed was mirrored in the individual growth of later forms. The concept of racial senility lying at the end of this pattern of growth may be a logical extension of the analogy between ontogeny and phylogeny, but it illustrates just how far this later version of recapitulation had come from the original progressionism. Haeckel's biogenetic law was the last flowering of the concept of unilinear progress, which had fascinated naturalists for centuries. The true impact of the Darwinian revolution can be seen in the fact that even the most fundamental opponents of the selection theory no longer accepted such a unified—or such an optimistically progressive—view of the relationship between embryology and evolution. Beyond that, perhaps it is possible to detect the influence of more than a scientific revolution. The growing willingness to believe that evolution might inevitably lead to degeneration seems to parallel the general loss of faith in progress that occurred at the end of the nineteenth century. Few of the scientists would have admitted it, but their concept of orthogenesis was more in tune with the pessimistic view of the cycle of human civilization expressed in Spengler's *Decline of the West*.[2]

Since orthogenesis represents a breakdown of the old progressionism, it might be fruitful to look for a link with the more general principle of the irreversibility of evolution stated in "Dollo's law." However, Louis Dollo's famous generalization merely expressed the belief that evolution is too complex a process ever to repeat itself exactly. It was sometimes interpreted as implying a kind of inevitability that must keep evolution driving along a particular direction, but this was never Dollo's own position.[3] Indeed, certain kinds of orthogenesis would definitely have contradicted Dollo's law. Racial senility was sometimes held to be the result of a retardation of growth returning a group to its original form. Such a view of evolution retracing its steps would have been incompatible with Dollo's (or any Darwinian's) view of irreversibility. Dollo explicitly repudiated the belief that degenerative evolution could return a form to its original state.[4] Even degeneration was an ongoing process that took the group onto new ground, although the exact nature of the new developments need not be predetermined by what went before.

A search for the motivations that prompted so many naturalists to adopt a theory of orthogenesis must concentrate on the two most important implications of the theory: the prevalence of nonadaptive characters and the linearity of evolution within each group. The belief that Darwinism exaggerated the importance of utility and adaptation became widespread at the turn of the century and was accepted by naturalists with diametrically opposed views on which alternative to seek. On one hand, the supporters of orthogenesis promoted large-scale trends under the control of some internal biological predisposition. On the other hand, there was also much support for the idea of salta-tive, or discontinuous, evolution, which in the form of De Vries's muta-tion theory postulated totally random genetic changes as the source of nonadaptive characters. The two alternatives were united only in the belief that Darwinism overstressed the extent to which the environment dominated the life of the individual and hence the course of evolution. Each insisted that internal changes within the organism could produce new forms able to survive, at least for a time, even if incompatible with the environment. At this level, acceptance of the widespread existence of nonadaptive characters was simply a reaction against the blanket utilitarianism of Darwin's theory, a feeling that it was going too far to assert that every character of every species that had ever inhabited the earth must have been shaped solely by adaptive considerations. It was also a reaction against the Darwinians' tendency to invent utilitarian ex-planations of characters they did not understand, with little hope of ever being able to confirm or refute the hypothesis. A generation of naturalists now decided to try out the possibility of explaining such characters in terms of purely biological forces, although they disagreed among themselves over the nature of these forces.

The remaining Darwinians wondered why so many naturalists were determined to repudiate one of Darwin's key insights. The danger of postulating developments independent of the environment was that all too easily they could take on a purpose of their own and could lead to a reintroduction of teleology, or finalism. Even De Vries recognized this danger and insisted that his mutations would eventually have to be judged by the environment. The long-range trends of orthogenesis posed a far greater problem, since in many cases they did indeed seem to be leading inevitably toward a predetermined goal. Yet the naturalists who supported the theory were not mystics determined to abandon all hope of obtaining a scientific explanation of evolution. They believed that they had firm evidence to back up their claim that Darwin's utilitarian-ism went too far. At the level of the small-scale characters used to dis-tinguish closely related species, it was often a matter of interpretation whether or not they were adaptive. Darwin and Wallace insisted that knowledge of the lifestyles of many species was so imperfect that scien-tists could not hope to discover the true purpose of many structures, while their opponents argued that some characters were obviously too trivial to be of any real value.[5] Such a debate was extremely difficult to resolve in practice, and it inevitably reflected the wider interests of the protagonists; just the same, the paleontologists believed that they could demonstrate the appearance of major nonadaptive features immediately before extinction. It required a good stretch of the imagination to think up a purpose for bizarre structures such as the enormous antlers of the Irish elk; and indeed, the reasonable nature of their claim is attested by the fact that many of the naturalists who contributed to the emergence of the modern synthesis felt it necessary to evade this question. To give just one example, Julian Huxley's concept of "allometry" was designed to show how the orthogenetic production of oversized characters could result as a by-product of natural selection. The emergence of the modern synthesis may have convinced most biologists that orthogenesis was not a significant factor in evolution, but until a really powerful Darwinian alternative was available there seemed no reason to give up the belief in nonadaptive characters.

The real distinguishing feature of orthogenesis was the claim that the internal factor in evolution was an orderly one producing linear developments over vast periods of time. Here the fossil evidence was less decisive. The Darwinians had always insisted that the fossil record was incomplete, thus obscuring many details of evolution; yet many later paleontologists were convinced that they could see the pattern clearly enough to recognize a degree of linearity incompatible with the selection theory. What motivation could have led them to ignore the possibility of concealed irregularities and to arrange the limited number of specimens in their possession into neat, regular patterns? In part, their approach may have stemmed from exactly the kind of overoptimism that had once fueled the belief in sudden episodes of creation. There is a natural tendency to assume that the available evidence is complete enough for significant analysis. To accept that most of what happened in the past is concealed by lack of evidence is to impose a limitation that many paleontologists have been unwilling to accept. Once it was agreed that the existing evidence was reasonably adequate, the temptation to arrange it into the simplest possible pattern would then appear to be the most scientific procedure.

This is an intriguing explanation, but almost certainly there was more to it than this. Why were so many paleontologists determined to ignore the realism of Lyell's and Darwin's views of the imperfection of the fossil record in their effort to find regular patterns of evolution? The scope of the support for orthogenesis suggests that the nature of their training leads biologists to look for order in nature, and that this can easily get out of hand. Darwin believed that however orderly the laws of nature, the material products of these laws were not arranged in a regular pattern. Evolution was thus a haphazard process with no consistency of purpose and no ultimate goal. For those naturalists who concentrated on morphology, on the study of biological forms without reference to the everyday requirements of life, it was difficult to accept such a view. They regarded formal relationships as the more important and were constantly tempted to look for order at this level. At one time, idealists such as Agassiz had postulated an overall harmony of development leading toward the goal of the human form. They had adopted a Platonic vision of an underlying order with a deeper significance than the superficial irregularities, and they believed that it was the true purpose of the naturalist to uncover this order. The developments of the late nineteenth century associated with the rise of Darwinism had convinced most naturalists that evolution was a branching process with no single goal. We have, however, already seen how the American school adapted Agassiz's idealism to the new situation by postulating regular patterns of evolution within each group modeled on the individual growth of the highest form in the group. The orthogenesis of Cope and Hyatt was thus derived from their fascination with the regularity of growth and development, a perspective which they brought with them from idealist philosophy. A similar link with idealism is apparent in Eimer's work. The origins of orthogenesis lay in an anti-Darwinian, antiutilitarian philosophy of nature in which formal morphological relationships took precedence over the demands of function and the environment.

It was this link with idealism that posed one of the greatest problems for the later supporters of orthogenesis. They were aware that the concept was rooted in a teleological view of nature, which the mechanistic attitude of later nineteenth-century science had declared out of date. The fact that Bergson used the evidence for orthogenesis to substantiate his claim that evolution was directed by a mysterious *élan vital* served as a vivid reminder of these origins. Few biologists now accepted a nonmaterialistic interpretation of orthogenesis, but they were constantly on the defensive, as though aware of the potentialities inherent in the idea. Several later supporters of orthogenesis deliberately set out to discount the link with teleology by arguing that a trend need not be supposed to have a goal drawing it onward. They stated that orthogenesis was the consequence of forces within the organism that predisposed it to vary in a certain way. The predisposition would arise from its physical or chemical constitution, not from any predetermined goal. The theory could thus be adapted to the new materialism, and even to Mendelian genetics, although the details of such a reconciliation proved impossible to work out to the satisfaction of either side.

If the later supporters of orthogenesis were not idealists, why did they continue to search for regular patterns in evolution even after this became unfashionable? Perhaps the only explanation is that the hypothesis of regular development exerted a fascination deeper than that of pure idealism. These naturalists expressed their opposition to Darwinism by claiming that the evolution of life could not be the result of chance. It must be based on natural law, and laws—as every scientist knows—impose order and regularity upon the processes of nature and prevent their unfolding in a haphazard manner. The primacy of law, rather than the Platonic goal of development, thus became the basis for this continuing opposition to Darwinism.

This argument was not, of course, a valid one. Darwinism does not deny that nature is governed by law, although Darwin and his followers were forced to admit that they did not understand all the laws involved with evolution. His theory is based on the belief that some natural processes are so complex, and involve so many interacting phenomena that the whole ensemble presents an appearance of irregularity and unpredictability, although each individual effect is governed by law. The orderliness of the laws themselves is only apparent in phenomena that are governed by a simple natural process, often at the microscopic level, such as in crystallization. Once we move onto a wider level of activity, too many effects interact, and the regularity is obscured. No one doubts that the earth's surface is shaped by natural, and hence lawlike, processes; but no one expects the mountains to have the shape of regular cones or pyramids. In the same way, evolution may be governed by laws, but it should not be expected to show linear patterns. Those naturalists who opposed Darwinism on the grounds that it denied the role of law had missed one of the central points of his argument. For all their attempts to search out mechanistic explanations, their "laws" were really nothing more than trends created by arranging limited evidence into artificially regular patterns. They mistook the products of their own imagination for an order built into the structure of nature. In this sense, even the later versions of orthogenesis can be seen as a vestigial influence of nineteenth-century idealism upon twentieth-century biology.

THE ORIGINS OF ORTHOGENESIS

Before the appearance of orthogenesis, the idea of regular, nonadaptive trends had established itself in a slightly different guise as one of the earliest alternatives to Darwinism. We saw in Chapter 3 how some of Darwin's opponents used the idealist version of the argument from design to defend natural theology. Theistic evolutionists such as Argyll, Carpenter, and Mivart argued that natural selection could not be an adequate explanation of evolution because there were patterns in nature that seemed to indicate a higher purpose than mere utility. They interpreted these patterns as evidence of a divine concern for order and beauty, which had somehow been built into the evolutionary process. Belief in an explicitly supernatural cause for such trends soon became unfashionable, but the trends themselves were the foundations upon which the later supporters of orthogenesis would build.

In Germany, the lingering remnants of idealism also ensured that there would be suspicion of any purely utilitarian mechanism of evolution, Darwinian or Lamarckian. Form was not solely dependent on function, but developed according to laws of its own—although the Germans were less likely than the British to invoke the direct hand of the Creator in establishing the laws. An indication of the way things would develop can be found in the writings of the botanist, Carl Nägeli. Already, in 1865, Nägeli had argued the widespread existence of non-adaptive characters and had forced Darwin to concede them as a major unexplained problem for his theory.[6] In 1884, his *Mechanisch-Physiologische Theorie der Abstammungslehre* expounded a theory of heredity and evolution that included a significant place for nonadaptive trends produced by natural predispositions within the germ plasm. The theory was not widely noted in the English-speaking world until an American translation of the book's concluding summary appeared in 1898. The appendix to this translation expressed surprise that a theory so closely allied to that of the American school had gone so long unnoticed and attributed this to the lack of adequate translation.[7]

Nägeli did not deny the role of adaptation in evolution, but he reduced it to an insignificant part. As chief driving force he introduced a "perfecting principle," or *Vervollkommnungskraft,* inherent in the material of heredity. Nägeli termed this material the "idioplasm," and like Weismann's germ plasm it was potentially immortal, being passed on from one generation to the next. Unlike the germ plasm, however, it resided in the whole organism, not in a single isolated location, thus allowing a quasi-Lamarckian explanation that adaptive changes were the result of continued direct action of the environment on the idioplasm. (Nägeli denied the inheritance of characters acquired solely during the life of the individual.)[8] New forms of life were constantly being created by spontaneous generation, each beginning a new line of progressive evolution. Nägeli, however, did not believe that there was a specific direction of progress: Evolution was a branching process, and each line of development was controlled by its own laws. The perfecting principle increased the level of complexity of the organisms, while permitting an ever greater variety of possible structures and hence an increasing tendency for evolution to differentiate into branches.[9] So far was Nägeli from creating a linear "chain of being" that he questioned the botanists' ability to rank the known forms of plants. The cause of the perfecting principle was thus not a teleological attraction toward a predetermined goal, but an organizing force, analogous to inertia, operating among the basic units of living matter (the micellae). The sheer complexity of possible interactions at this level ensured the open-ended nature of the progression.[10]

Although Nägeli stressed the role of an internal force working independently of the environment, this was not a theory of true orthogenesis. Despite his insistence on the limited role of adaptation, his perfecting principle was utilitarian in the sense that it increased the level of complexity and constantly opened up new evolutionary opportunities. He postulated mechanical forces within the idioplasm to explain the principle and resisted any charge that his theory reintroduced a mystical or teleological element. Indeed, it was precisely the lack of any fixed direction of progress that allowed him to sustain this position: A perfecting principle that did go in a predetermined direction would certainly have seemed teleological. Even so, the similarities between Nägeli's perfecting principle and Bergson's later, but explicitly vitalist, *élan vital* are obvious enough. Many biologists of the time refused to accept Nägeli's excuses and regarded his theory with suspicion. The concept of unqualified orthogenesis emerged when naturalists who were equally unconvinced by Darwin's utilitarianism began to invoke internal forces that produce highly structured trends, while at the same time they tried to evade the charge of reintroducing teleology by denying that the trends had any progressive tendency.

Wilhelm Haacke's introduction of the term "orthogenesis" came in an 1893 discussion of heredity and variation in which he proposed that the germ plasm consists of geometrically structured elements whose rearrangements are limited by their actual shape.[11] This created a structural predisposition to vary in a particular direction, which would account for linear, nonadaptive evolution. Curiously, Theodor Eimer succeeded in popularizing the concept of orthogenesis and in giving it scientific respectability precisely by denying that it was the result of a solely internal predisposition. He criticized Nägeli for postulating a mystical driving force and constantly maintained that in a truly mechanistic theory such as his own, an external stimulus would be needed to elicit the directed variation.[12] This argument appears to have been successful in convincing at least some scientists that orthogenesis did not imply teleology[13]—although it is evident from a study of Eimer's views that the actual course of orthogenetic evolution was completely predetermined by the internal predisposition to vary in a particular direction.

Eimer's first important biological work was a study of the wall lizards of Capri and was published in 1874.[14] In it he argued that a new variety or subspecies could be produced by sudden mutation, without reference to the demands of the environment. At this point, however, he was prepared to concede that only those varieties with adaptive characters would be able to establish themselves permanently, and he was actually criticized for implying that the coloration of the Capri lizards was adaptive. As Weismann's germ plasm theory was elaborated in the course of the 1880s, Eimer became increasingly less tolerant of the selection theory. He soon emerged as one of Weismann's most vitriolic critics, constantly grumbling that his own views were being suppressed by the influence of his powerful opponent. His *Entstehung der Arten* (1888), translated into English as *Organic Evolution,* supported a Lamarckian explanation of adaptive characters;[15] but Eimer had not forgotten his work on the wall lizard, and he was now becoming convinced that directed variation could produce major nonadaptive trends in evolution.

The evidence that Eimer used to substantiate this claim was derived from a study of the colors and patterns on the wings of butterflies. An address outlining this work to the Leiden Congress of Zoology in 1895 was translated into English[16] and formed the first chapter of his *Orthogenesis der Schmetterlinge.* The nature of his subject matter forced Eimer to launch an attack against a Darwinian stronghold, the theory of protective coloration and mimicry. Since an insect's colors are not under its voluntary control, the Darwinians had assumed that protective coloration must have been developed by selection rather than use-inheritance. Eimer boldly set out to challenge the view that the colors had any adaptive value at all. He insisted that many cases of camouflage were merely products of the naturalists' imagination and pointed out that no amount of Darwinian rhetoric would explain the colors on the inside of a snail's shell.[17] The coloration of animals in general was controlled by internal factors directing variation along particular lines, not by the environment. Eimer even maintained that mimicry in insects was due not to selection but to unrelated species following the same basic law of color-variation.[18]

Eimer's evidence depended upon recognizing an evolutionary connection between various species in the genus *Papilio.* The basic sequence of evolution was assumed to be from those forms with longitudinal stripes through to spots, cross-stripes, and finally uniform color. Although he was convinced that he could distinguish which were the primitive and which the more highly evolved patterns, Eimer could give no adequate justification for his claims. Since there was no time factor involved—they were all living species—his work suffered from the grave disadvantage that the patterns of evolution could be dismissed as mere figments of his imagination, created by arranging the wide variety of modern forms into a totally artificial framework based on the simplest possible linear arrangement. A Darwinian would require a good deal of convincing that any modern species could have exactly preserved the character of an ancestral form while others continued to evolve. The weakness of this kind of argument suggests why much of the later emphasis on orthogenesis switched to paleontology, where a sequence in time was, at least in theory, observable.

Eimer claimed that "the causes of definitely directed evolution are contained . . . in the effects produced by outward circumstances and influences upon the constitution of a given organism."[19] His insistence upon a role for external causes was, as we have seen, a device for checkmating the claim that the course of evolution was in some mysterious way predetermined by the original constitution of living things. It also allowed Eimer to call upon the Lamarckians' evidence for a direct effect of external conditions upon the germ plasm, including Standfuss's experiments on the inherited effects of temperature upon coloration in butterflies.[20] He could also account for a limited amount of evolutionary divergence by noting that the same species in different areas might be affected by different stimuli.[21] At the same time, however, there are passages in Eimer's writings that make clear his intention of reducing the evolution of color in the animal kingdom to a few distinct trends, which would have to be internally predetermined for them to have this degree of consistency. The evolution of the butterflies consisted of many parallel lines running independently through the same pattern of development, and similar patterns could be seen in other groups, including even the vertebrates. The same rigid limits on variation must hold for every other character in both the animal and vegetable kingdoms: "Orthogenesis is a universal law. It holds, as I have long insisted, not only for the markings, but also for the other morphological characters of animals, and also for those of plants. Even in the latter, as my personal observations have convinced me, the markings of the blossoms and the shape of the leaves follow, in rigorous conformity to law, a *few* definite directions."[22] The "laws" of orthogenesis had obviously become something more than interactions between the constitution of organisms and external stimuli. The evolution of life had been able to unfold along certain paths and no others, with divergence only being possible within the combinations allowed by these fixed trends.

The motivation that drove Eimer to postulate so orderly a development of life was seldom made explicit in his writings, but there are clues that show us he was, in fact, influenced by the lingering remnants of the old idealist *Naturphilosophie*. These influences are exactly the same as those transmitted to the New World by Agassiz to become incorporated into the orthogenetic Lamarckism of the American school. Eimer too accepted the recapitulation theory as a link between the laws of individual development and evolution. He also believed in the discontinuity of evolution, seeing the orthogenetic trend as broken up into a series of distinct stages. His writings also contain an exaggerated eulogy for the principle of the unity of nature and a favorable reference to the philosophy of Lorenz Oken.[23] For all his efforts to present orthogenesis as a mechanistic theory, its real purpose was still the imposition of a transcendental order upon the apparent chaos of nature.

The personal tone of Eimer's attacks upon Weismann upset even some potential supporters and allowed Weismann to dismiss him as a bigoted enthusiast, but there is evidence that Weismann was forced to modify his own views in response to the challenge. Eimer always liked to point out that some of Weismann's earliest researches actually supported the possibility of directed variation. Weismann's *Studien zur Descendenztheorie* (1875) reported observations on butterflies and caterpillars that were aimed, not at eliminating this possibility, but at reducing it to a scope such that it could not play a major role in evolution. In his earlier discussions of the germ plasm theory, he admitted that under certain circumstances external influences might provoke directed rather than random variation.[24] By using this device, the neo-Darwinists could explain away cases of the apparent inheritance of nonadaptive characters without admitting the Lamarckians' claim that more positive changes in the body could be inherited. Nevertheless, Weismann soon seems to have realized that this idea dangerously compromised his principle of the isolation of the germ plasm, and he sought instead to explain directed variation by a mechanism completely internal to the germ itself.

This was the process of germinal selection, according to which variation could be directed by the competition for nourishment among the determinants, or character-units, of the germ plasm.[25] Weismann now held that it was possible for variation itself to move in the direction favored by selection, since the selection process would pick out those individuals whose germ plasm contained a strong determinant for the appropriate character. Apart from producing this character to benefit the individual, the determinant would increase its power to affect the next generation through its success in the germinal struggle for existence. At first, Weismann stressed the negative effects of germinal selection as a means of speeding up the elimination of characters that had become useless, but by 1902 he had conceded that there might also be a random factor that would occasionally allow a useless determinant to achieve dominance. In this case, the resulting variation-trend would cause nonadaptive orthogenesis. However, there was no structure within the germ plasm that could impose consistent trends of the kind postulated by Eimer. At best, germinal selection would allow for a few cases of overdevelopment, not vast lines of parallel evolution affecting whole groups.[26]

Although Weismann always tried to reduce directed variation to a trivial level, germinal selection represented a major concession to his opponents, who greeted it as an illustration of the purely speculative nature of his entire theory. There were many who shared Eimer's belief that orthogenesis ought to play a more prominent role in evolution than even germinal selection would allow, although few took the idea quite as far as he did. One of Eimer's leading supporters was the British marine biologist Joseph T. Cunningham, whose translation of *Die Entstehung der Arten* appeared in 1890. As we have already seen, Cunningham was a prominent Lamarckian, but he also believed that there were nonadaptive characters in all species, and he had challenged Wallace on this point. In 1895 he postulated an internal force producing nonadaptive variation in the flatfish he was studying. In addition to the Lamarckian factor there was

> a tendency to definite variation, or growth in different directions, leading to a manifold variety of regular, definite symmetrical forms. This tendency can only be regarded as internal to the organism, as connected with the tendency to growth and multiplication inherent in organic units. . . . Whatever the causes of non-adaptive variation, the resulting structural features are the regular "geometrical" forms and characters which the multitude of different organic forms present in such marvellous diversity.[27]

Although Cunningham was prepared to arrange the various species of flatfish into a regular evolutionary sequence based on their markings, he limited his trend to within a single group. He thus avoided the implausibility of Eimer's trends, which were supposed to affect the entire animal kingdom. It was the character of each form that determined the possibilities of regular variation, although Cunningham was never able to specify details of how the predispositions would be imposed. His suggestion does, however, reveal the fascination that some naturalists felt in the possibility of seeing linear patterns linking the forms they studied. In Cunningham's case the fascination was short-lived, though, and he eventually went on to develop his Lamarckian theory based on the role of hormones in heredity.

Eimer died in 1898, and one could be forgiven for assuming that interest soon switched to paleontology as a source of evidence for orthogenesis. Certainly, there was a basic weakness in the method of arranging living forms into sequences that were only hypothetically equivalent to a scale of evolution. The fossil record offered a way out of this dilemma, and there were many paleontologists who became advocates of orthogenesis. Nevertheless, it would be wrong to assume that Eimer's approach was abandoned: In a later study, Glenn L. Jepsen recorded a number of twentieth-century authors who had published evidence for orthogenesis based on living forms.[28] There were few attempts to demonstrate trends on the scale postulated by Eimer, but at a more restricted level it was still argued that the laws of individual growth predetermined the direction of variation and produced non-adaptive evolution. What exactly were these "laws of growth," and how were they imposed on the organism and hence on the species? The development of a more materialistic approach to biological theory put increasing pressure on the supporters of orthogenesis to come up with a new kind of answer to these questions.

In the idealist tradition, the laws of growth were defined in a purely morphological sense. A particular pattern of development was assumed to be somehow imprinted upon the individual, ensuring that its form could only unfold in a certain way. Orthogenetic evolution occurred when latent stages of the growth pattern were manifested through additional growth in new individuals. By the end of the nineteenth century, it was no longer fashionable to think in terms of purely morphological laws, since this was too reminiscent of the idealists' explicit appeal to a transcendental source of order in development. A morphological series was merely a trend, not a law in its own right, and it needed to be explained as the result of physicochemical laws affecting the structure of the individual. Even Eimer and Cunningham had paid lip service to this requirement, although their laws were still specified in morphological terms. There was, however, still a need to provide an actual mechanism to explain how the organism was predisposed to vary in a certain direction, and it was becoming ever more pressing with the contemporary advances in embryology and the theory of heredity. A few naturalists ignored the new fashion: D'Arcy Wentworth Thompson is the best example, his work admired as a tour de force but dismissed as a theoretical anachronism. To remain viable in this new atmosphere, the supporters of orthogenesis would at least have to appear to offer mechanistic explanations, even if the details could never be worked out in a manner that would satisfy the laboratory biologist.

There were, however, two quite different techniques that could be used to specify the causes limiting variation. The one that would seem most obvious to the modern reader would require the postulation of forces predisposing the germ plasm to vary in a certain way; in effect, a built-in tendency of the genes to mutate only in one direction. If it were assumed that the genetic information was stored in a chemical form, then in the absence of any detailed knowledge of the structures involved, the possibility of only limited molecular rearrangements could not be ruled out. A few geneticists seem to have toyed with this idea, but in general the supporters of orthogenesis ignored it. Like the Lamarckians, they refused to accept the dogma of the isolated germ plasm and continued to think of heredity and the growth process as an integrated system with a two-way flow of information between the material of heredity and the growing body. They preferred to imagine that the predisposition to vary was built into the growth process itself, not into the material carrying the hereditary information. If there were chemical or physical factors in the growth of the individual that might tend to cause additional growth in some direction, this might accumulate over many generations to become the cause of orthogenesis. Such a viewpoint, of course, still required the Lamarckian principle of the inheritance of acquired characters—although in this case the characters were acquired not in response to an external challenge but as a result of internal growth tendencies. Here again we see the intimate link between Lamarckism and orthogenesis, one based on the common assumption that individual growth played a role in evolution, and we can appreciate why both failed to adjust to the rise of modern genetics. In a sense, the experimental study of the growth process, which the Germans called *Entwickelungsmechanik,* was a blind alley for evolution theory. Whatever its success in embryology, it allowed a generation of biologists to continue with the belief that by studying ontogeny they were making a genuine contribution to scientific evolution theory. Kammerer's Lamarckism and most theories of orthogenesis fell into exactly the same trap of assuming that because one could postulate, or even demonstrate, a modification of individual growth, this must somehow be reflected in heredity and evolution. In only one way was this approach able to go beyond the old morphological laws of growth: Instead of requiring absolutely predetermined patterns of evolution that almost begged for a teleological interpretation, it allowed one to think in terms of material processes that merely limited the range of possible variation. This offered a more plausible materialistic interpretation and allowed orthogenesis to adapt its image, at least superficially, to the twentieth century.

One of the last biologists to think in terms of pure morphology was D'Arcy Wentworth Thompson, whose noteworthy *On Growth and Form* appeared in 1917. The purpose of this book was to suggest that a mathematical analysis could be used to show the extent to which organic forms were molded by the requirements of physical law. Much of the work was compatible, if somewhat out of tune, with the new breed of materialism; but at the end of the book Thompson introduced mathematical relations between forms, which were distinctly idealist in tone and which had definite implications for evolution theory. Although not really interested in the details of the evolutionary mechanism, Thompson's anti-Darwinian bias was evident. He argued that similarities of form were not enough to guarantee an evolutionary relationship between two species because the purely formal laws of growth might produce the same structure in different circumstances.[29] His famous demonstration that the range of structures within a class can be derived from a single form by changing the geometrical coordinates against which it is displayed provided evidence for this. Thompson believed that these geometrical relationships indicated how the forces shaping growth must have controlled the evolution of the various forms. "If . . . diverse and dissimilar fishes can be referred as a whole to identical functions of very different coordinate systems, this fact will itself constitute a proof that variation has proceeded on definite and orderly lines, that a comprehensive "law of growth" has pervaded the whole structure in its integrity, and that some more or less simple and recognizable system of forces has been in control"[30] The implication of this seems to be that the laws of mechanics are somehow capable of imposing patterns upon growth, an idea that might not have seemed implausible in individual cases, but that could hardly have been expected to apply to relationships on the scale postulated. If the laws were powerful enough to control the evolution of an entire class, then the forces of nature were imposing a degree of unity more reminiscent of the idealist view of morphological relationships.

Thompson believed that physical forces imposed strict limits upon the way organisms could grow and evolve. The alternative was to suppose that chemical forces working within the matter of which the organisms are composed can have the same control over growth. This was the approach suggested by the Russian naturalist Leo S. Berg, whose

Nomogenesis appeared in 1922 and was translated into English four years later—with an introduction by Thompson. Berg used all lines of evidence, including the fossil record, to establish the claim that evolution had been an orderly process. He was totally opposed to the view that chance played a role in variation and was determined to show that all aspects of evolution were governed by law.[31] External causes were not relevant—witness the fact that the great ice ages had had little overall effect on the direction of evolution.[32] Instead, it was internal forces that were constantly driving variation along rigidly determined lines. Berg even insisted that many separate lines of evolution had advanced in parallel through the same sequence of vertebrate classes leading up to the mammals.[33] This went far beyond Thompson's claim, returning almost to the old unilinear progressionism. The basic course of evolution consisted of the unfolding of preexisting rudimentary structures under the influence of internal forces. These forces were inherent in the chemical structure of living matter. "That there are intrinsic and constitutional agencies laid down in the chemical structure of the protoplasm, which compel the organism to vary in a *determined direction,* may be inferred from the fact that not infrequently evolution proceeds, as it were, in face of the environment, in a direction leading the organism to destruction."[34] Such a statement appears quite materialistic; yet in general Berg would not agree that orthogenesis was a purposeless force. He believed that most evolution was necessarily progressive and led to the regular appearance of more specialized characters. This was as Theodosius Dobzhansky put it, the Achilles heel of his theory, for to suppose that protoplasm was created with all these potentialities inherent within it was necessarily to invoke a purpose imposed on nature from without.[35] Although Berg provided a masterly summing up of the evidence for orderly evolution, he betrayed the cause of orthogenesis by reviving the link with progress that had carefully been purged by Eimer.

Thompson and Berg spoiled the case for orthogenesis by exaggerating the scope of the relationships to the extent that their hints at control by physical forces seemed quite implausible. Just the same, they had sketched two interesting possibilities as to how variation might be limited. Less well-known writers were already following up both approaches on a smaller scale and constructing more acceptable interpretations of orthogenesis. Thompson's view that mechanical forces were responsible for shaping growth had its roots in the morphological tradition and was the more difficult to exploit. Nevertheless, it was possible to argue that purely physical constraints upon the growth of the individual might become cumulative and affect evolution. This was suggested by Bashford Dean in connection with the structure of the egg cases in the chimaeroid fishes.[36] Far more plausible in most cases was Berg's assumption that chemical forces directed growth and variation, although the suggestion that protoplasm might be predetermined to change only in certain ways was too primitive to be of any use. The expansion of biochemistry and its application to embryology created a new situation in which it became possible to argue that chemicals released during the growth of the individual might alter the adult form and thus direct evolution. A tendency to overproduce a growth hormone, for instance, might be the cause of orthogenesis and the evolution of bizarre structures. This approach was advocated by the biochemist L. J. Henderson at a symposium on orthogenesis held by the American Society of Zoologists in 1921.[37] Other biochemists do not seem to have been eager to follow up the suggestion, but we shall see below that many of the paleontologists searching for an explanation of orthogenesis would find it a fertile source of speculative inspiration.

The overenthusiasm of writers such as Thompson and Berg made it all the more necessary for scientists in closer touch with twentieth-century attitudes to proclaim the emancipation of orthogenesis from teleology. One of the most sophisticated discussions of this question was by the American naturalist Charles Otis Whitman, the first director of the Woods Hole Marine Biological Laboratory. Whitman was sympathetic to Darwinism on the grounds that it had freed biology from teleology. His deep interest in embryology made him suspicious of the early, oversimplified versions of Mendelism and the mutation theory and convinced him that evolution occurred through the accumulation of minute variations. A study of color variation among pigeons, published posthumously in 1919, had convinced him that variation was not always random, but could sometimes move preferentially in certain directions. Like Eimer, he believed that he could detect evolutionary relationships among living species, which were paralleled in the ontogeny of the more advanced forms as predicted by the recapitulation theory. But Eimer's concept of evolution rigidly determined by law had led many biologists to believe that orthogenesis was teleological. Whitman's own suggestion was that the nature of individual growth imposed limits on the possible range of future variation, thereby eliminating all hints of a goal toward which the process must be drawn.

If a designer sets limits to variation in order to reach a definite end, the direction of events is teleological; but if organization and the laws of development exclude some lines of variation and favor others, there is certainly nothing supernatural in this, and nothing which is incompatible with natural selection. Natural selection may enter at any stage of orthogenetic variation, preserve and modify in various directions the results over which it may have had no previous control.[38]

By limiting the scope of orthogenesis, Whitman thus held out the hope of a reconciliation with Darwinism—a rare example of flexibility in an age of rigidly polarized opinions.

Whitman's position has been hailed by Ernst Mayr as an important anticipation of the modern view that the existing genotype does indeed impose some limit on the range of possible variation.[39] Whitman would probably have allowed orthogenesis a greater role in evolution than any modern biologist, but it is certainly true that the concept of limits on variation was far more plausible than the morphologists' rigid patterns of evolution. Whitman described germinal variation as the only source of new characters and was opposed to Lamarckism on the grounds that it required a flow of information in the reverse direction from the fundamental process of ontogeny. This would suggest that he was indeed thinking in terms of what today would be called a genetic limit to variability—except that the passage quoted above still mentions the traditional view that there are "laws of development" controlling variation. This makes it difficult to be sure that he had completely rejected the possibility of ontogeny playing a constructive role in evolution. Although he was no slavish adherent of the recapitulation theory,[40] nevertheless he did accept it, and hence accepted the belief that variation consists of additions to growth. While denying the possibility of an external influence being transmitted back to the germ plasm, he may still have been prepared to agree that a character acquired by an internal extension to growth might be transmitted by heredity. The paleontologists who were now making up the front line defense of orthogenesis certainly believed that internally stimulated modifications of ontogeny could provide a nongenetic source of orthogenetic variation. For all the sophistication of Whitman's hypothesis that orthogenesis works by limiting rather than by actually directing variation, one cannot be certain that he would have rejected their ideas.

ORTHOGENESIS AND PALEONTOLOGY

Paleontology generated the most optimistic hopes of conclusive evidence for orthogenesis because only the fossil record could reveal the actual course of evolution over a long period of time and confirm the existence of parallel linear trends. In addition, paleontology enjoyed a number of qualities that inclined its practitioners toward an orthogenetic viewpoint. I have already referred to the temptation to regard the existing state of the record as complete enough for detailed analysis. This allowed paleontologists to ignore the possibility that a number of side branches might be hidden by lack of information and enabled them to arrange the few specimens available to them into the simplest and hence most regular genealogies. They were also dealing with extinct forms, which provided great opportunities for dismissing unusual characters as useless or even harmful. Since one cannot observe the lifestyle of an extinct animal, it was easy to imagine that any enlarged or apparently bizarre structure was nonadaptive and hence a sign of racial senility.

Paleontologists always looked at evolution from a distance, so to speak. They could describe a change, but could not investigate its causes directly, and may thus have been inclined to raise a purely descriptive trend to the status of an explanatory "law." They accepted loose analogies with totally unrelated phenomena such as inertia or crystallization as though they were valid contributions to evolution theory. Even when pressured into suggesting a fairly detailed mechanism of evolution, the paleontologist was not concerned if the hypothesis bore little relation to the theories of those biologists studying variation on a day-to-day level. Some paleontologists actually suggested that trends spread over a geological time scale might forever lie beyond the scope of laboratory investigation. Support for orthogenesis was upheld by all of these factors, so that in the end paleontology became a discipline apart, separated from other branches of biology by a wall of suspicion and indifference. Only with the rise of a powerful new initiative in these other areas were the paleontologists at last forced to consider whether or not their evidence was open to reinterpretation.

The paleontologists' concept of racial senility contrasted sharply with Eimer's view of orthogenesis. His trends were nonadaptive in the sense that they were indifferent to the demands of the environment and could be studied in purely morphological terms. That a trend might actually run against the requirements of adaptation was a disturbing thought; but this extension of the theory could be interpreted in two different ways. First, for many invertebrate paleontologists, following the lead of Hyatt, it was simply an extension of the idealist view of linear evolution modeled on the pattern of embryological growth. Racial senility came when the group had run out of steam, and it represented an evolutionary retreat back to a state resembling that from which it began. There was degeneration in the sense of a loss of advanced characters acquired during the group's heyday, a trend that was a symptom rather than the actual cause of the decline toward extinction.

Many vertebrate paleontologists, however, adopted a different outlook based on the overdevelopment of progressive characters. An adaptive trend was supposed to acquire a momentum of its own, which eventually drove beyond the limits of utility, and produced excessively large or bizarre structures that interfered with the animal's lifestyle. Of the two approaches, it was this which more easily emancipated itself from the old idealist philosophy and offered the best hope of being translated into a mechanistic theory. Indeed, it represented a significant breakdown of the attempt to understand evolution solely in terms of morphological patterns. Overdevelopment seemed almost to demand an explanation based on a disturbance of the growth process, pushing paleontologists to reinterpret the link that was supposed to exist between ontogeny and phylogeny. It also required one to think very carefully about the role of adaptation, instead of leaving the scientist free to ignore the question of utility. It would be necessary to define just how far a trend might go before the animals became unable to cope. There would also be the extremely delicate task of postulating a mechanism that would allow the environment to create adaptive trends, and yet would give those trends a degree of autonomy in the control of evolution.

It was no accident that some of the first attempts to create a theory of what would later be known as orthogenesis were made by paleontologists studying extinct cephalopods, especially the ammonites. These creatures had enjoyed a remarkable evolutionary career of progress and expansion, followed by degeneration and extinction, all clearly recorded in their fossilized shells. Since the individual grew by adding new compartments to its shell, its entire life was displayed in the fossil, thus permitting comparisons to be made between individual growth and evolution. Not only was the group's progress toward more complex forms enshrined in the ontogeny of later individuals, but the eventual degeneration was anticipated by their senile characters. Here was an orderly pattern of progress and decline, apparently based on the rhythm of individual growth, senility, and death. The pattern was strange enough to puzzle many a convinced Darwinian and gave plenty of scope for those who were in any case predisposed toward a morphological "law" of evolution.

The parallels between the life history of the individual and the group in the ammonoids had been pointed out earlier in the century by Alcide d'Orbigny and others. In the 1860s, two paleontologists began to develop the evolutionary implications of this relationship and created what was, in effect, a theory of orthogenesis. Of the two, the lesser known is Wilhelm Waagen, a German naturalist who spent his later career on the staff of the Geological Survey of India. In an important paper written in 1869, Waagen introduced the term "mutation" to denote the regular pattern of evolution he found in some ammonites.[41] Here the term had a meaning quite different from that later adopted by De Vries and subsequently incorporated into modern genetics. Waagen's mutations were not sudden changes, but gradual ones, their distinguishing character being that they proceeded consistently in a single direction. In a later commentary on his Indian work, Waagen noted that the pattern of evolution there was the same as in Europe—but since the rocks were different, it could not have been the external conditions that were responsible for the changes. Evolution thus occurred because of a "tendency of organisms to produce an offspring varying in a certain and defined direction."[42] The sequences were clear enough to be predictive, allowing the experienced paleontologist to know in advance what kind of descendants would result from a given form. Waagen was convinced that similar fundamental laws governed the evolution of other groups, but since he refused to elaborate on the nature of the forces involved, his views were not widely influential.

The concept of regular evolution running toward racial senility was pioneered by the American paleontologist Alpheus Hyatt. We have already seen how the influence of Agassiz led Hyatt to develop the embryological analogy for the ammonoids in 1866. Even then he was insisting on old age as a distinct phase in both the growth of the individual and the evolution of the group, but as yet he had offered no explanation of why evolution consisted of the regular addition of stages to growth.[43] Eventually he accounted for the earlier, more positive phase of development in Lamarckian terms, invoking use-inheritance under the influence of a particular lifestyle to explain the consistent and parallel developments within the group; but this still left unanswered the question of racial old age, which could hardly be seen as a consequence of use and habit. Hyatt was convinced that a steady degeneration was the prelude to extinction. Forms that had become progressively more coiled and ornamented gradually uncoiled and simplified themselves in the geological eras preceding their final disappearance. On one point he was firm: The degeneration was not due to a retardation of growth that simply returned the form to an exact equivalent of its primitive ancestor.[44] This had been suggested by Hyatt's colleague E. D. Cope to allow for degeneration by the loss of growth-stages, thus making evolution reversible. Hyatt did not accept retardation and continued to insist that degeneration was due to the addition of new stages to growth. If that were so, why, then, should the additions resemble the simple ancestral forms?

Hyatt believed that there was a limit beyond which no further development of a group could take place. Given favorable conditions, it could progress to this limit, but would be unable to cope with any further changes in the environment. He had noticed a similarity between senile forms and pathological ones and had realized that if an individual were exposed to unfavorable conditions it would develop a pathological structure, whereas if the group were exposed to such conditions it would degenerate as a whole. The appearance of senile types followed a general change in the earth's conditions with which the group could not cope, coming to a climax in the Cretaceous period when the ammonoids became extinct.[45] The pathological effects became ever more pronounced as the adverse conditions continued because the effect on each generation was absorbed into the race through the acceleration of growth—in effect, there had been inheritance of acquired nonadaptive characters. Although the changed conditions were the stimulus that provoked the degeneration, however, there was a sense in which they were not its true cause. The group could no longer cope because it had run out of evolutionary energy, just as the individual organism eventually reached the limit of its powers. Although there was no predetermined return to the original form, the main feature of the pathological reaction was a loss of the more complex characters that had evolved in earlier periods. All species undergoing degeneration thus tended to follow a similar pattern back to a simple form resembling the original starting point. If there was no absolute law directing the variation, the interaction between the adverse conditions and the group's internal constitution imposed well-defined limits upon the degeneration.

At the turn of the century, a group of American invertebrate paleontologists followed the lead given by Hyatt. Charles E. Beecher and John M. Clarke extended the technique of relating ontogeny and phylogeny to the Brachiopoda, Robert T. Jackson to the Pelecypoda and Echinoidea, and Amadeus Grabau to the Gastropoda, while James P. Smith continued with the ammonites.[46] All of these workers agreed that the stages of individual growth could be used to classify the sequence of forms defining the pattern of evolution. All acknowledged that the progressive response of the organism to its environment was the cause of orderly, linear evolution, but admitted a senile phase after the group's powers of adaptation were exhausted. Just the same, the concept of an entire group running out of evolutionary energy smacked a little too much of vitalism, and the idea did not remain popular in the new century. Paleontologists began looking for clues that would point the way to a mechanistic explanation of racial senility, which would be more in tune with developments elsewhere in biology. Even among the invertebrates there were some cases in which the prelude to extinction was not degeneration but an overdevelopment of progressive characters. A favorite example was the overcoiling of the shell of the mollusk *Gryphaea,* which was supposed to have continued to such an extent that in the end the creatures could hardly open their shells.[47] One of Hyatt's pupils, Charles E. Beecher, suggested that the progressive development of spines and other ornaments was the result of an inbuilt tendency toward the production of excess shell material. If some natural explanation of the trend toward the overdevelopment of shells could be found, orthogenesis would move firmly into the twentieth century.

The best attempt at such a theory was made by W. D. Lang of the British Museum (Natural History), who tackled the problem of overdevelopment in his introduction to the *Cataloque of the Fossil Bryozoa* in 1921. The Bryozoa are minute creatures living in colonies, and it was to the colonies themselves, rather than to the individual organisms, that Lang and others found they could apply the recapitulation theory. Lang was convinced that much parallel evolution had taken place in this group, as elsewhere. The basis of his theory was a claim that the shells of all invertebrates are the by-products of an uncontrolled tendency of the organism to produce calcium carbonate. Haeckel had once suggested that the chemical nature of this substance must affect the way it is laid down to form shells, and Beecher had postulated a compulsive tendency to build more complex structures. Lang now declared that calcium carbonate was the great determining factor of invertebrate evolution. He did not deny the ability of natural selection to ensure that the hard material was laid down in the manner best suited to protect the organism, but this was not the primary origin of the shell.[48] The organism's metabolism forced it to excrete calcium carbonate, and over many generations the extent of the material secreted built up until it reached a stage where it could no longer be usefully employed and began to cause inconvenience. At first there was an increasing elaboration of structure as the material built up, but in the end the sheer thickness of the shell might produce the pathological simplification seen by Hyatt. In the case of the Bryozoa, eventually "the amount of calcium carbonate in the skeleton becomes so pronounced that, not only does the skeletal structure become secondarily simple by the piling up of calcareous matter, so further evolution is inconceivable, but the very life-processes of the organism appear to be in danger of obstruction, so constricted and tunnel-like become all the apertures in the skeleton, by which the organism communicates with its environment."[49] The limited possibilities of disposing of the excess calcium carbonate accounted for the regularity of evolution characteristic of orthogenesis, which Lang believed gave a degree of predictability to evolution.

The key to Lang's theory was his mechanism for explaining the overproduction of calcium carbonate. He suggested that all shells were originally built of chitin, a material secreted in order to eliminate waste nitrogenous matter. However, at some point in time a disturbance of the organisms' metabolism had converted the secretion to calcium carbonate. At first the tendency to produce this material had been successfully held in check by inhibiting factors, but these in turn had been removed by further disturbances of the metabolism. It was possible that environmental changes were responsible for triggering these disturbances, thus accounting for the rapid bursts of evolution that Cope had called "expression points." The end result was always the same, however; all barriers to the production of calcium carbonate were removed, the metabolism was totally given over to this useless task, and the organism was literally stifled to death by its own waste products.

Lang knew that his theory was incompatible with Mendelian genetics, which he dismissed as a "mere manifestation of mongrelism."[50] However, he was able to exploit an analogy with William Bateson's emphasis on the prevalence of inhibitor genes, which could be removed by destructive mutation to reveal the character they had once masked.[51] Nevertheless, Lang wanted to use the term "potentiality" in a different way from the Mendelians. He did not accept the existence of a structure in the cell nucleus with the potentiality to produce a given character. On the contrary, for Lang a potentiality was a much wider function of the organism as a whole that predisposed it toward orthogenetic variation. In this sense his theory was still a continuation of the nineteenth-century belief that processes operating in the growth of the individual hold the key to evolution. It may be significant that when he tried to deal with the regularity of the patterns in which the waste matter was laid down, he could only suggest that it

> points . . . to compulsion from within—to an inherent tendency in the ancestral form which becomes actual as its evolution is worked out in the offspring, although these are distributed among many divergent lineages. Such a potentiality in a radical form, becoming actual many times over in its descendants, introduces an element of inevitability in evolution, and makes it possible to speak of the

evolutionary aim of a lineage, without thereby implying anything of what is usually understood by teleology.[52]

Despite the disclaimer about teleology, it is possible to see this aspect of Lang's theory as a direct continuation of earlier attempts to unify evolution into predetermined, regular patterns.

Lang's approach was thus a curious amalgam of the old and the new. It still contained a relic of the old idealist concept of development, but in a deliberate attempt to appear modern he had thrown the emphasis onto his mechanistic explanation of the basic overproduction. Racial senility was no longer defined as the end product of a morphological trend, but was explained in terms of biochemical disturbances to ontogeny. There were few references to Hyatt in Lang's work, but several to the vertebrate paleontologists who were supporting overdevelopment, along with a hint that excess bone might be produced by a mechanism parallel to that which he had described for calcium carbonate. Even more than Lang, the vertebrate paleontologists were trying to play down the notion of a preordained path of development, concentrating instead on less structured trends in which overdevelopment could be explained in biochemical terms. Their ideas still avoided the essential points of Mendelian genetics, but at least the superficial trappings of twentieth-century science could be displayed.

The possibility of nonadaptive trends in invertebrate evolution had been recognized since at least the 1860s, but among vertebrate paleontologists such ideas do not seem to have become popular until the end of the century. In America, Cope had stressed the linearity of evolution, but had concentrated on adaptive trends that could be explained by Lamarckism and the consistent effects of habit. Cope had never denied the existence of nonadaptive characters, but this had not been a leading feature of his views on evolution. By 1900, however, the situation was changing as nonadaptive trends became more widely postulated. Now the emphasis was not on the creation of imaginary patterns relating species, but on individual cases in which a particularly bizarre structure seemed to have gone beyond the bounds of utility. There was a failure of the imagination, an unwillingness to admit that extinct forms might have had lifestyles so different from those of any living animals that those grotesque structures might have had a purpose scientists could not understand. At first sight, this approach seemed to offer an even more obvious challenge to the logic of Darwinism. The environment was supposed to exert so little pressure that a species could evolve structures that were not only useless but actually harmful.

Only when it became impossible for any form of livelihood to be sustained did extinction occur. In this sense, of course, the idea was radically un-Darwinian, although it could be made to appear consistent with mechanistic biology as, for example, in Lang's theory. Yet unlike Lang, the vertebrate paleontologists were forced to think carefully about the relationship between orthogenesis and adaptation. Lang postulated that the shells of invertebrates were entirely the products of a nonadaptive trend, and were only secondarily used for the purpose of protection. In the case of overdeveloped horns and teeth, however, the structures were evidently first evolved for use and only later pushed byond the limit of utility. It was thus obvious that at some point in the process the influence of the environment must have been involved in the creation of the orthogenetic trend.

If the new concern for overdevelopment somewhat paradoxically refocused attention on the relationship between the organism and its environment, there remained the much wider question of the supposed linearity of evolution. Some of the most important paleontologists supporting orthogenesis did not accept the overdevelopment theory of extinction at all. Henry Fairfield Osborn, for instance, admitted that overspecialization might lead to extinction by making the species less flexible in the face of environmental challenge—but this was a long way short of nonadaptive overdevelopment. Yet Osborn continued the traditions of the American school by insisting that there were linear trends in evolution that could not be explained by natural selection. He saw whole groups of species moving in parallel through the same trend, just as Cope had done, although he rejected habit as the guiding force and looked for some internal biological compulsion to vary. The linearity of evolution was meant to be anti-Darwinian, even if the overall direction was not grossly out of step with adaptation. The classic display of horse evolution set up in 1910 by the American Museum of Natural History, which certainly stressed the linear view of evolution favored by Osborn, was an attempt to popularize orthogenesis, even though the trend was described as an adaptation to a new way of life. The paleontologists of the twentieth century were more aware of the subtle relationship between the internal and external control of evolution than their idealist forbears, but they were still convinced that variation must be a lawlike process capable of ordering evolution in a way that Darwinism ignored.

Since paleontologists had argued that variation was directed rather than random, it was now more than ever necessary that they propose a mechanism by which the process was controlled. Jepsen's claim that for most paleontologists orthogenesis was simply a term describing the existence of trends cannot be accepted.[53] There were some who merely described the trends and expressed a general acceptance of linear evolution; but there were others who developed hypothetical mechanisms to explain the direction of variation, equivalent to Lang's theory of invertebrate orthogenesis. Osborn himself devoted a great deal of thought to this question, and although he eventually gave up and fell back on the claim that the trends were too slow to show up in the laboratory, he was aware of how unsatisfactory this position must seem to others. However firmly convinced the paleontologists were of the evidence for their trends, they knew that no one else would take them seriously unless they could come up with a plausible mechanism. As A. F. Shull put it in 1934, "What the world needs, then, is not a good five cent cigar, but a workable—and correct theory of orthogenesis."[54]

The problem was that like the cigar, such a theory was hard to find. Indeed, the laboratory biologists would have said that most paleontologists were looking in the wrong place, since their theories accepted nongenetic modifications of ontogeny as the cause of directed variation. Like the Lamarckians, they assumed that the cumulative effect of—in this case, nonadaptive—additions to growth could shape the course of evolution. Most popular was Lang's approach based on biochemical disturbances of the growth process that perhaps were triggered by environmental changes; but whatever support this approach might derive from a few biochemists, it still evaded the central problem of how the changes became imprinted on the germ plasm. If one tried to think instead of genetic predispositions to vary in a certain direction, there was then the problem of explaining how the predisposition was at first set up to create a potentially adaptive trend. This implied a role for the environment, and hence either a parallel with Lamarckism or a return to teleology. The paleontologists' support for orthogenesis all too often foundered on exactly the same problem as Lamarckism: the impossibility of explaining an interaction between the organism and its environment in terms of the new genetics. Orthogenesis was stronger than Lamarckism during the 1920s and 30s only because its "acquired characters" seemed a less artificial modification of the organism's growth, and because they were supported by fossil evidence suggesting a very slowly developing, and hence perhaps indirect, connection between the environment and the genes.

Overdevelopment was first suggested as a cause of extinction by Ludwig Döderlein in 1888 and was then taken up by Ernst Koken in 1902, F. B. Loomis in 1905, and Charles Depéret in 1907.[55] Classic examples were the excessive size and decoration of some dinosaurs, the teeth of the saber-toothed tiger (which were thought to have become so long that the animal could not eat), and the antlers of the so-called Irish elk (which became so large it could no longer hold its head up).[56] Depéret presented perhaps the clearest illustration of the theory's implications when he extended it to include some modern forms.

> Each of these lines culminates sooner or later in mutations of great size and highly specialized characters, which become extinct and leave no descendants. When one line disappears by extinction it hands the torch, so to speak, to another line which has hitherto evolved more slowly, and this line in its turn traverses the phases of maturity and old age which lead it inevitably to its doom. The species and genera of the present day belong to lines that have not yet reached the senile phase: but it may be surmised that some of them, e.g. elephants, whales and ostriches, are approaching the final phase of their existence.[57]

Lamarckism and the consistent effects of habit were clearly not enough to sustain this chilling image of entire groups being driven inexorably toward their destruction. It was necessary to invoke some biological mechanism outside the organism's voluntary control to explain this degree of nonadaptive variation.

In 1911 the British Lamarckian Arthur Dendy suggested a way in which the inheritance of acquired nonadaptive characters could account for this "momentum of evolution" leading to overdevelopment and extinction.[58] Since growth was controlled by hormones, he argued that natural selection would act to favor those individuals who had hormones that promoted the growth of useful organs. It would also eliminate any glands producing secretions that checked the useful growth; but once selection had succeeded in producing individuals with no such checks in their metabolism, the way was opened for unrestricted growth, since the checks could never be restored by further variation. There would now be nothing to stop the once useful organ from growing beyond the limits of utility. Slight excesses of the appropriate hormone would mean significant increases in growth, which would accumulate over the generations. Dendy insisted that such increases would be inherited if they occurred before the germ cells of the individual had matured—a clear link with Lamarckism. Since the extra variation had occurred by addition to growth, it would be recapitulated in the ontogeny of the later individuals. These suggestions were subsequently cited by Lang as an important influence upon his own thinking about invertebrate overdevelopment.

A rather different, but still essentially Lamarckian, explanation was offered for certain long-range trends in the ancient Amphibia, first described by the British paleontologist D.M.S. Watson in 1919 and 1925. Watson showed that several important characters of the skeleton had undergone regular, parallel development in a number of related amphibians. Since in the course of these developments the creatures had evolved from an aquatic to a terrestrial and then back to an aquatic mode of life, the changes were not adaptive. In fact, Watson himself did not believe that any environmental influence could be the cause, suggesting instead that the changes must "owe their constancy of direction to the mechanism which determines the characters of the adult being so constituted as to be capable of modification only in certain definite ways."[59] This could be taken to imply a genetic predisposition, although Watson did not elaborate on the causal mechanism. In 1930, however, Baron Francis Nopsca pointed out that a nonadaptive effect of the environment could be used to explain at least the later phases of these trends. He argued that the amphibians' return to an aquatic mode of life was directly responsible for the changes in the skeleton, since the water affected the thyroid gland and thereby altered the ratio of bone to cartilage.[60] For this mechanism to be effective over the time period covered by Watson's trends, one would have to assume the cumulative inheritance of acquired nonadaptive characters.

An interesting variation on the same theme was the concept of racial disease suggested by Richard Swan Lull. In his frequently reprinted textbook, *Organic Evolution,* Lull noted Cope's evidence for linear kinetogenesis, but pointed out the lack of experimental evidence for use-inheritance. He cited the wide range of evidence for the linearity of adaptive evolution and for nonadaptive trends, but offered no detailed mechanism to explain how either was controlled. In 1924, in a more specialized discussion of dinosaur evolution, Lull stated that the driving force was, in general, adaptation to changing conditions and that the final extinction of the group was due to a geological revolution. In addition, however, he did raise the prospect that "grotesques, like *Stegosaurus,* were from their very nature doomed to a speedy extinction as a result of the mortal outcome of racial disease."[61] The riotous growth of some structures may have been stimulated by overactivity of the pituitary gland, itself caused by a lack of chemicals such as iodine in the diet. Here was a purely environmental cause of overdevelopment, although Lull carefully refrained from implying that there would be a cumulative inheritance of the effect. Presumably each generation suffered greater problems due to a progressive scarcity of vital nutrients, until eventually the whole species succumbed to a disease affecting all the individual animals.

All of these mechanisms attributed orthogenesis to a nongenetic source of variation, with the exception of Watson's hint at a built-in predisposition independent of the environment. This latter alternative had in part been anticipated by Weismann's germinal selection, although this was never taken very seriously and was not, in any case, applicable to really long-range trends. Watson was clearly thinking of some actual structure in the group's germinal constitution that would require variation only in one direction. A similar suggestion had been made somewhat earlier by another British paleontologist, A. Smith Woodward, who postulated "some inherent property in living things, which is as definite as that of crystallization in inorganic substances."[62] By itself, this sounds no more sensible than Berg's idea of a chemical predisposition of all organic matter to progress; but Smith Woodward did at least insist that the predisposition lay within the germ plasm, and he stressed its nonadaptive consequences. The problem with his suggestion, as with others, was that by making the effect so general, it failed to explain why the trends manifested themselves in so many different ways among the various kinds of animals. A satisfactory theory of orthogenesis would have to show how each particular trend became built into the constitution of a particular group of animals, which implies that it would have to take into account the circumstances under which the group evolved. The difficulty of developing such a theory was increased by the fact that the trend had to emerge out of what was originally an adaptive phase of evolution. Did this mean that adaptation actually helped to set up the variation-trend, and if so, how did the trend subsequently gain the power to set off on its own?

One possibility was that natural selection itself participated in setting up the trend by picking out those germ plasms with a tendency to vary in the required direction, thus engendering the momentum that eventually led to overdevelopment. Dendy had proposed something along these lines, although his trends had been nongenetic, but in general the supporters of orthogenesis were unwilling to credit selection with such an active part. One reason for this was the application of orthogenesis to another long-standing problem with Darwinism: the supposed inability of selection to explain how organs developed in their incipient stages before they became large enough to be of any use. F. B. Loomis's frequently cited paper on "Momentum in Variation" concentrated on examples of overdevelopment, but also hinted that a variation-trend independent of adaptation would explain these early stages of development.[63] Another American, Maynard Metcalf, expanded this suggestion into a major argument for orthogenesis.[64] If this point were accepted, selection could not by itself create the momentum of variation. The germ plasm would have to acquire the tendency to vary in a certain direction from some other cause, which would still leave the question of how it was able to anticipate the future requirements of adaptation.

Loomis's article seemed to imply that the origin of the variation-trends was essentially random—an equally unknown equivalent of Darwin's random individual variation. Trends in many different directions might spontaneously arise within a species from time to time; by chance, a few would coincide with the species' future needs and would be aided by selection until they went too far and gave rise to overdevelopment. Some would create adaptively neutral characters and be totally useless. Others would definitely be harmful, but might nevertheless be able to flourish for a while. Loomis conceded that a really dangerous trend would soon be extinguished by selection, but he suggested that a lesser "handicap" might develop for some time until it became so obstructive that it caused extinction. Such a position implied that evolution enjoyed a good deal of freedom from environmental control, but a somewhat similar view of the spontaneous origin of variation-trends could be adopted by naturalists prepared to give the environment a rather more active say in the matter.

There were many supporters of orthogenesis who did not believe that the trends could get very far out of step with the requirements of adaptation. For them, the evidence for a non-Darwinian factor that controlled variation was the linearity of the trends rather than any non-adaptive consequences. W. D. Matthew, for instance, whose analysis of horse evolution was cited as favorable to orthogenesis, attacked the theory of extinction from overdevelopment in the case of the fangs of the saber-toothed tiger.[65] Perhaps better known is William Berryman Scott, who had originally supported the neo-Lamarckism of Cope, but who abandoned this explanation of the fossil trends in the last decade of the nineteenth century. In his article "On Variations and Mutations" he emphasized the evidence for parallel linear evolution against the Darwinian theory based on random individual variations. He adopted Waagen's definition of mutation and insisted that such variation-trends did not imply some mystical directing force. For the rest of his career, Scott continued to advocate linear evolution, yet he was never attracted to the overdevelopment theory. While admitting the difficulties of explaining how the saber-toothed tiger used its jaws, he regarded it as "quite certain that no arrangement which was disadvantageous, or even inefficient, could have persisted for such vast periods of time."[66] In effect, the variation-trends were under the strict control of natural selection: "Mutations are the effect of dynamical agencies acting long in a uniform way and the results controlled by natural selection."[67] Precisely what these dynamical agencies were, Scott consistently refused to speculate, although the phrase implies something more active than a purely germinal predisposition to vary. Whatever the source of the trends, they could only succeed so long as they were consistent with the demands of the environment.

It was the partial correlation between the trends and the demands of adaptation that led Scott's friend Henry Fairfield Osborn to suspect that there must be some input from the environment in the creation of what subsequently became an internal or totally germinal predisposition to vary. Originally a Lamarckian, Osborn too had abandoned this position during the 1890s, switching at first to the mechanism of organic selection. He soon decided, however, that the consistency of orthogenetic variation demanded a biological, rather than a merely behavioral, determining factor in evolution. Osborn's own fossil evidence convinced him that linear trends usually began in an adaptive direction—indeed, they anticipated adaptation by producing a structure that would only become useful when fully developed. Once established, however, the trend went on inexorably enlarging the structure in a manner that was not compatible with the selection of random variations. The best illustration of this came from Osborn's study of the Titanotheres, an extinct order of mammals, especially from characters such as the horns. He believed that the paelontologist should study the development of single characters, without bothering to divide the evolutionary line arbitrarily into species.[68] By concentrating on a single character such as the horns, it could then be shown that in members of a group such as the Titanotheres there was a tendency for the character to show the same pattern of development. The horn always appeared at the same point on the skull and then grew steadily in size, although different lines within the group might pass through the pattern at different rates. Osborn called such linear trends "rectigradations."[69] Eventually, the character thus produced became useful to the animals, but the trend continued whether further increases in size were useful or not. In the case of the Titanotheres' horns, they eventually became extremely cumbersome and may have contributed to the group's extinction.[70] Once a definitely inadaptive stage was reached, the line was driven to extinction by natural selection and was replaced by another line that had not advanced so far in the pattern.

The rectigradations were not built into the essential structure of organic matter, since each trend was peculiar to its own group. It was Osborn who pioneered the term "adaptive radiation" to denote the rapid diversification of a newly dominant class into a wide range of different forms that exploited the various ecological opportunities open to them.[71] Only when the orders within the class were firmly established as distinct branches did linear trends become apparent. It was therefore necessary to imagine a process that created the trend and imposed it upon the germ plasm of the group at this early stage in its history. The trend could not arise spontaneously in the group, since its timing and its direction seemed to owe something to the demands of the environment. Yet Osborn could not adopt a simple Lamarckian position, if only because he was increasingly aware of the lack of any experimental evidence for so direct a link between environment and germ plasm. He proposed that evolution must be a process of "tetrakinesis," or the interaction of four distinct factors arbitrated by natural selection: the germ, the development of the individual, and the physical and organic environments.[72] The inclusion of ontogeny as a controlling factor in evolution illustrates the link between Osborn's views and the earlier traditions of orthogenesis and Lamarckism, while at the same time he was clearly trying to make his position more acceptable in the light of modern knowledge. The crucial question remaining was essentially that posed by Lamarckism, but rephrased in a somewhat more sophisticated form: How did the environment and the soma influence the germ plasm so as to set up a variation-trend controlling later evolution?

Unlike many paleontologists, Osborn refused to retreat into isolation. He was aware of the new discoveries in the study of heredity and did his best to synthesize his own views with those of the laboratory biologists. His *Origin and Evolution of Life* (1917) presented the fruits of his thoughts along these lines; yet in the end he had to admit that he could see no satisfactory way of explaining how genetic trends were set up. Perhaps biochemical catalysts such as enzymes were somehow able to influence the germ plasm;[73] but this was only a hint, and in the absence of any more detailed theory Osborn was forced to insist on the validity of his own data. Rectigradations did occur, even if their mechanism was not revealed by laboratory investigations. Perhaps such long-range trends occurred too slowly for their effects to be appreciable under artificial conditions. Later on he became even more explicit:

Although these new characters arise independently in different phyla, thus pointing to some kind of germinal predetermination or predisposition, they do not indicate an internal perfecting tendency, because they are timed with reference to external-internal reactions. Consequently there appears to be some kind of interaction between environment, habit, and the time of appearance of these new organs;

but we have no inkling as to what this relation is—whether, in fact, it is causal unless it consists of some kind of physico-chemical interaction.[74]

With the hint that the relationship might not be causal at all, Osborn came close to admitting that his search for a mechanism of adaptive orthogenesis had broken down, leaving him face to face with the age-old alternative of teleology. The same admission is apparent in the name "aristogenesis" that he introduced in 1932 to denote what he was convinced was the essentially goal-directed process of linear evolution.[75]

Osborn's suggestion that the trends might not be accessible to laboratory study led Thomas Hunt Morgan privately to accuse him of toying with mysticism.[76] This was not quite fair, since Osborn was doing his best at the time to work out a mechanism that would have seemed plausible to Morgan and the other geneticists. Nevertheless, his approach was doomed from the start, given the preconceptions of the new science. Genetics was essentially hereditarian in outlook; it held that the genes alone controlled the growth of the organism and that evolution could only occur through changes arising spontaneously within the material structure of the genes. It might have been possible to argue that purely nonadaptive trends could be created by a chemical predisposition of the gene to mutate more often in one direction than any other—indeed, we shall see that some biologists did explore this avenue, without much success. To require, however, that the trends be shaped by the environment was going too far, since it reintroduced the bogey of Lamarckism in a more elusive form than ever. If Osborn dissociated himself from orthodox Lamarckism, he was left with trends that anticipated the requirements of adaptation, but that were set up by a totally unverifiable process. It was hardly surprising that a laboratory biologist would regard this as a sellout to the old form of teleology. Thus, the most ambitious attempt to create a theory of orthogenesis foundered on the fact that the trends were no longer considered to be totally separate from utilitarian concerns. So long as linear evolution was totally nonadaptive, as Eimer intended, it could always be argued that there was an unknown genetic mechanism accounting for the trends; but once a link with adaptation was introduced, the cause was lost. Even if the trends had enough genetic inertia to produce over-developed structures, their origin required an input from the environment that was unacceptable to twentieth-century geneticists.

REACTIONS TO ORTHOGENESIS

Many of the laboratory biologists who supported Mendelism and the mutation theory were themselves convinced that adaptation played a very restricted role in evolution. At the same time, they wanted nothing to do with artificially regular morphological patterns and preferred to believe that mutation was a random process caused by some unknown chemical rearrangement of the germ plasm. A few efforts were, however, made to provide evidence that mutations might occur more readily in certain directions, particularly under the influence of an external stimulus. Vernon Kellogg found a group of peculiarly marked beetles on the campus of Stanford University, and he interpreted this as a case of directed variation.[77] There were also laboratory demonstrations of environmental changes stimulating nonadaptive variation in a single direction. Standfuss's experiments on heat-induced modifications in butterflies were frequently cited by the Lamarckians, but were appropriated with rather more justification by Eimer. The problem was, however, that directed variation elicited by a particular environmental stimulus could hardly be expected to produce consistent trends of the kind postulated by the more extreme supporters of orthogenesis.

The most widely discussed effort to link this approach with modern genetics was the work of Richard Goldschmidt and V. Jollos during the 1930s. Jollos published evidence that exposure to heat precipitated certain kinds of mutation in the fruit fly, *Drosophila*. This was supported in America by Goldschmidt, although counterevidence was soon reported and the effect was eventually forgotten.[78] A. F. Shull—who, it will be remembered, wanted a theory of orthogenesis more than a five-cent cigar—expressed the hope that directed mutations would turn out to be the answer, but was prudent enough to report the lack of confirmation for Jollos's work.[79] It is doubtful whether such an effect could ever have fulfilled Shull's wish for a workable theory, since there was no indication that the species would keep mutating in the same direction whatever the external circumstances, which is what the paleontologists were trying to demonstrate. Goldschmidt, however, continued to keep a place for orthogenesis within his highly unorthodox genetics. Unlike most geneticists, he retained a strong interest in the orderliness of the growth process and believed that we would only see the outcome of those mutations that produced an effect consistent with the existing condition. Unlike Whitman, he was prepared to see that this limited viable mutations to a single direction.

The development of any primordium is closely interwoven with that of all other primordia and, therefore, a local change caused by a mutation affecting an early embryonic process . . . cannot lead to a viable result if the embryo is not able to carry out the proper regulations. The selection of the direction in which genetic change may push the organism is therefore not left to the action of the environment upon the organism, but is controlled by the surroundings of the primordium in ontogeny, by the possibility of changing the ontogenetic process without destroying the whole fabric of development . . . Thus what is called in a general way the mechanics of development will decide the direction of possible evolutionary changes. In many cases there will be only one direction. This is orthogenesis without Lamarckism, without mysticism, without selection of adult conditions. [80]

Goldschmidt remained an important critic of modern genetics, but his views did not gain wide acceptance. The genetical theory of natural selection accepted only the most general kinds of limits imposed on the possibility of variation by the existing genotype.

The paleontologists' conception of orderly development was also based on a (less sophisticated) belief that ontogeny guides evolution along particular lines. Yet they had actually created the evidence for this belief out of the available fossil record. The collapse of orthogenesis in the face of the modern synthesis was engineered by paleontologists such as George Gaylord Simpson, whose *Tempo and Mode in Evolution* (1944) showed that the fossil record was in fact compatible with the newly revived Darwinism. [81] Simpson argued that most cases of supposedly linear evolution were based upon oversimplified arrangements of inadequate evidence. From the modern perspective, it would be easy to dismiss the supporters of orthogenesis as visionaries trying to impose a hoped-for order upon the chaos of nature by a deliberate act of the imagination. Quite clearly they were spurred on by an explicitly anti-Darwinian, or even pre-Darwinian, philosophy of nature. One cannot, however, dismiss their work as unscientific on these grounds, and the best reason why not is the fact that the originators of the modern synthesis were forced to make a number of concessions to the evidence for orthogenesis, especially on the question of nonadaptive characters. Of course they were able to dismiss some cases of supposed overdevelopment as illusory. J.B.S. Haldane also argued that intraspecific competition, such as occurs in sexual selection, can produce bizarre characters that enable the organism to reproduce, but which may be harmful in a wider context. Although this explains nonadaptive characters in terms of the selection theory, it concedes a basic point that had always been implicit in the theory of sexual selection: that the demands of the environment do not impose absolute control upon evolution. Haldane also acknowledged that no satisfactory explanation existed of the kind of trends uncovered by Hyatt's school—although invertebrate paleontologists were already making efforts to account for these developments in terms of adaption to changing habits.[82]

Perhaps the most significant concession on the question of non-adaptive trends was Julian Huxley's mechanism of allometry. In a sense this exploited a suggestion made long ago by Darwin himself, who had used the term "correlation of growth" to denote the linkage of two or more characters in an organism's constitution. Even before 1900, E. Ray Lankester had urged this phenomenon as an explanation of nonadaptive evolution, since if a useless character was correlated with a useful one, selection would have to promote both together.[83] This was subsequently translated into the language of genetics by Thomas Hunt Morgan, who noted that a single gene often had an effect on more than one character.[84] A. H. Sturtevant and Julian Huxley expanded the idea by including the phenomenon of relative growth rates. Sturtevant argued that if the antlers of the Irish elk, for instance, were correlated with testicular secretions, then selection for the latter because of reproductive superiority would also increase the size of the antlers.[85] Huxley tackled the more complex case of Osborn's Titanotheres by noting that the growth rate for horns might be different from that of the body as a whole.[86] If selection boosted sheer size as a survival factor, the horns might be forced to increase an at even greater rate and might ultimately reach cumbersome proportions. In this case, orthogenesis was not rejected, but was instead incorporated into the modern synthesis as a by-product of selection.

If the Darwinists were forced to explain away some cases of non-adaptive evolution, they were obviously opposed to the whole idea of unifying the history of life into a few mysterious trends. The danger that orthogenesis would lead to a reintroduction of teleology had never been far from anyone's mind, despite the frantic efforts to come up with a plausible mechanistic theory. At the 1920 meeting of the British Association, the invertebrate paleontologist Francis A. Bather launched what was probably the most wide-ranging attack on the philosophy of the movement. He was contemptuous of the lack of practical experience that led some paleontologists to deny the utility of many characters; but far more serious in its implications was the constant temptation to see a nonphysical cause as the driving force of regular trends, a danger that Bather emphasized by pointing to Bergson's incorporation of orthogenesis into his ostensibly vitalist philosophy. The analogy between ontogeny and phylogeny was singled out for particular criticism because as a source of teleological thinking it constantly tempted the biologist to think in terms of "some transcendental assumption, some unknown entelechy that starts and controls the machine, but must forever evade the methods of our science."[87]

In conclusion, Bather briefly referred to the human implications of orthogenesis. Not surprisingly, few supporters of the racial senility theory had cared to dwell upon what would happen if it were expanded into a general philosophy of evolution; but it is evident that by allying itself with this concept, orthogenesis had moved from the realm of pure idealism into a thoroughgoing pessimism. The implications were that all species—the human included—were foredoomed to pass through an inevitable pattern of progress and decline. Such a view would have been an outlook on humanity's destiny far more hopeless than anything attributed to Darwinism. Most supporters of orthogenesis refused to carry it this far. Osborn, for instance, carefully limited his explanation of human evolution to the process of organic selection, in which conscious choice of habits guided specialization without requiring a Lamarckian effect.[88] Yet there were a few examples of orthogenesis being applied to human beings—enough to show that Bather's fears were not unfounded. Hyatt had persistently maintained that certain human characters were "phylogerontic," and he had argued against the emancipation of women on the grounds that this would speed up the elimination of sexual differences that was one aspect of racial senility.[89] As late as 1933 another American, George W. Crile, attributed certain human disabilities such as the tendency toward peptic ulcers to the effects of orthogenesis.[90] Speaking long before Crile wrote his article, Bather pointed out the implications of this approach. Natural selection had been resisted because it destroyed the vision of inevitable progress and left living things at the mercy of their environment. Paradoxically, in an effort to maintain the autonomy of internal biological forces, the supporters of orthogenesis had created a nightmare. Far better, urged Bather, to abandon the inevitability of progress and accept that it could only occur through our own exertions.

> If we are to accept the principle of predetermination, or of blind growth-force, we must accept also a check on our efforts to improve breeds, including those of man, by any other means than crossing, and elimination of unfit strains. In spite of all that we may do in this way, there remain those decadent races, whether ostriches or human beings, which "await alike the inevitable hour." If, on the other hand, we adopt the view that the life-history of

races is a response to their past environment, then it follows, no doubt, that the past history of living creatures will have been determined by conditions outside their control, it follows that the idea of human progress as a biological law ceases to be tenable; but since man has the power of altering his environment and of adapting racial characters through conscious selection, it also follows that progress will not of necessity be followed by decadence; rather that, by aiming at a high mark, by deepening our knowledge of ourselves and our world, and by controlling our energy and guiding our efforts in the light of that knowledge, we may prolong and hasten our ascent to ages and heights as yet beyond prophetic vision.[91]

8

The Mutation Theory

T HE alternatives to Darwinism that I have explored so far all originated in a longstanding tradition that organic development must be an orderly process controlled by laws inherent in life itself. Once the original form of Darwinism had exhausted its immediate creative potential, a reaction set in that allowed naturalists to express more clearly than before their deep-rooted hostility to the oversimplified image of life at the mercy of capricious external forces. Traditional beliefs that had temporarily been rejected because they seemed too obviously teleological when measured against Darwinism's air of scientific materialism were rehabilitated and turned into equally "scientific" alternatives to selection that nevertheless retained key aspects of the older way of thought. Lamarckism drew upon the notion of the individual's creative response to environmental challenge in order to promote a more optimistic and purposeful view of adaptation. Orthogenesis eliminated the response to the environment and used the concept of addition to growth as a means of imposing a formal order upon the evolutionary process. These alternatives were successful for a time, not only because they appealed to an older philosophy of life that many naturalists had still not advanced beyond, but also because Darwinism seemed moribund and incapable of furthering biological research. Now, however, both Darwinism and the alternatives were to be challenged by a new approach to the study of what had emerged as the most crucial problems: variation and heredity. Fed up with the techniques of morphology and field study, which now seemed to have reached the limits of their usefulness, a new generation of biologists emerged, convinced that these problems would yield to the experimental method. Instead of setting up unverifiable hypotheses about the course of evolution or the nature of the germ plasm, they would look in detail at the facts of heredity and

[182]

variation in their search for generalizations that might ultimately be built into a new and more scientific theory of evolution.

This movement led biologists to recognize the significance of Mendel's laws and to introduce the concept of genetic mutation. These would prove to be vital factors in the eventual reinvigoration of Darwinism, although at first genetics and the mutation theory appeared to be contributing to its eclipse. The new approach was essentially hereditarian in outlook—committed to what had been Weismann's dogma of the isolated germ plasm. It was thus hostile to Lamarckism and potentially compatible with the selection theory. Nevertheless, its supporters could not follow Weismann in his support for Darwinism, both because their techniques inevitably favored an interest in easily detectable discontinuous variations and because the artificially isolated atmosphere of the laboratory fostered contempt for the field naturalists' intuitive feelings about the dependence of the organism upon its environment. It was thus possible for Hugo De Vries's mutation theory to postulate the sudden creation of new species without reference to selection. Internal rather than external factors once again became the key to evolution; but unlike orthogenesis, the mutation theory contained nothing predisposing it toward rigidly guided genetic change. The biometrical school—the chief bastion of Darwinism remaining in Britain—did not help matters when it dismissed the investigation of discontinuous variation as a wild-goose chase. Their preference for continuous variation symbolized their effort to adapt the new precision of investigation to the most fruitful insight of Darwinism: its insistence that evolution be treated as a complex process occurring within a population, rather than as a summation of individal changes. This hostility to the idea of discontinuous genetic change, coupled with the lack of interest in population thinking that characterized the first, very basic investigations of Mendelian heredity, ensured that a science of population genetics would be hard to create. The synthesis did eventually occur, of course—and when it did, it laid the foundations of a new Darwinism; but it took the skill of a few pioneers and the willingness to compromise of a new generation of biologists to overcome the bitterness of the original dispute.

In part, Darwinism was saved from extinction by the fact that the new science of heredity was just as implacably opposed to the other alternatives. Instead of combining against the selection theory, the new generation of opponents fought with and eventually destroyed the old. The rigid hereditarianism of geneticists determined to see the characters they investigated as inviolable biological units counted heavily against their ever accepting Lamarckism. Had the experimental evidence been unequivocal, they might have been forced to acknowledge that genetic units could be modified by external influences. Instead, the geneticists knew that the evidence for Lamarckism was of dubious quality, and their predisposition toward hereditarianism soon led them to search for ways of explaining away each supposedly Lamarkian effect. For the most part, geneticists were also hostile to the philosophy of development underlying the recapitulation theory. Their interest lay not in the process of ontogeny, but in the identification of particular somatic characters with hypothetical genetic units. They believed that variation could occur only through the recombination of such units or through the creation of new ones by mutation, not by the gradual addition of stages to individual growth. There was thus no reason to suppose that ontogeny recapitulated phylogeny and every reason to suppose that ontogeny did not guide evolution.

Lamarckism was overwhelmed not just because it lacked experimental support, but also because it could not find a suitably mechanistic replacement for the theory of variation by addition to growth. Orthogenesis was in a similar position, since even the biochemical theories of overdevelopment ignored the central hereditarian assumption of genetics. Yet there was a possible link between the mutation theory and nonadaptive orthogenesis, since both were based on the assumption that evolution must be under the control of internal rather than external factors. There was no necessary reason why mutations should occur preferentially in a single direction; but given the lack of understanding of the chemical structures that stored genetic information, the possibility could not be ruled out. If the evidence suggested it, the supporters of the mutation theory were willing to consider directed variation, and a few of them did take it seriously. In general, however, they were more concerned with the autonomy of the genetic processes and were quite happy with the idea of random (that is, inconsistent) variation. Even so, the proponents of both orthogenesis and the mutation theory agreed on one thing: that they preferred evolution by law rather than by chance. Even if the mutations were not consistently directed, they occurred as the result of causal changes within the germ plasm. Evolution thus followed the biological laws of mutation and was not subject to capricious and unpredictable changes in the environment. No one seemed to recognize that evolution was now reduced to the "chance" of whatever mutations happened to appear at the time, and that the mutation theory thus presented an image of totally undirected evolution and lacked even the sense of purpose that had been imposed by Darwin's emphasis on adaptation. Because the changes were in principle thought to be understandable in purely biological terms, the theory was hailed as a triumph of law over chance.

The potential for a link between the philosophy of orthogenesis and mutationism was symptomatic of a deeper level of continuity between the new and the old alternatives to Darwinism. At the most obvious level, genetics and indeed the whole experimental biology movement were opposed to the positive role that the more traditional anti-Darwinian theories gave to ontogeny. This is why it proved so difficult to find new interpretations of Lamarckism and orthogenesis that would satisfy both the mechanists and the exponents of the recapitulation theory. The old, more teleological view of development was necessarily destroyed by the new hereditarianism, although most of the latter's basic principles were not fundamentally opposed to the view of heredity adopted by Darwin and Weismann. In this sense, the initial opposition of the Mendelians to the selection theory was only skin-deep; it was founded on personal feelings generated by the trivial debate over the level of discontinuity in variation, as well as on the overenthusiasm of the laboratory students of heredity, which led them to ignore the possibility that their discoveries might be subject to the pressure of the environment in the wild state. It was inevitable that such disagreements would eventually be sorted out and a synthesis reached, with the only casualty being the earlier type of antiselectionist philosophy. The story, however, is not quite as simple as this because there was another strand of thought, which for a while helped to interweave the old and the new. The new wave of opposition to Darwinism was not always quite so radical as it now seems in retrospect, and it is important to distinguish the more conservative elements that at one time linked at least some of its concepts with the existing anti-Darwinian movement.

The possibility of a genetic explanation for orthogenesis was one such link, but it was by no means the only one, nor the most fundamental. The belief that evolution must result from law rather than chance was more basic, but there were also specific aspects of some early Mendelian and mutationist theories that served as attempts to synthesize the new and the old. Before he discovered the experimental evidence for the undirected nature of mutations, Thomas Hunt Morgan was fascinated by the idea that a series of linked mutations might produce linear nonadaptive evolution. The possibility that evolution might be no more than the unfolding of certain potentials built into the original forms of life was suggested by William Bateson and by J. P. Lotsy in his theory of the origin of new types by hybridization. Bateson also invoked the role of physical forces in ontogeny as a means of imposing harmonic regularities upon the discontinuous steps in variation. Although most field naturalists and paleontologists refused to accept the new emphasis on discontinuity, a few took up the idea and extended it to include the notion of a period of large-scale mutations in the distant past as a means of accounting for the origin of phyla and classes. Again there was an implication that life had gradually used up the fixed amount of evolutionary energy with which it was originally endowed. All this suggests that the theories favored by the experimentalists—and some of the experimentalists themselves—were by no means as free from the traditional ways of looking at nature as is sometimes imagined. It took time, as well as the extremely convincing kind of experimental evidence for the true nature of mutations that was provided by certain key organisms such as *Drosophila,* to create a way of thinking about evolution that was completely free of the old teleology. The chief purpose of this chapter is to reveal the extent to which the apparently new wave of opposition to Darwinism drew some of its emotional appeal and its antiselectionist arguments from the philosophy that already existed as a more fundamental alternative—in addition to the really new but rather superficial points of difference that were eventually resolved. The emergence of modern genetics and the genetical theory of natural selection represents not only the triumph of mechanism over the teleological view of development, but also the successful purging of certain less obvious aspects of the old way of thought from genetics itself.

DISCONTINUOUS EVOLUTION AND MENDELISM

Because both Mendelism and Darwinism contributed to the modern synthesis, the story of their estrangement and reconciliation has been told before and need not be repeated in detail here. Most attention, however, has focused on the "modern" aspects of the geneticists' thought, even when those aspects were pushed to an extreme that generated a superficial anti-Darwinism. The ideas that can be linked with earlier philosophies of evolution have not received the same level of attention. We shall see how the prominent English geneticist William Bateson drew much of his antiselectionist inspiration from ideas that were not incorporated into mainstream genetics and that reveal his concern for issues more reminiscent of the philosophy underlying orthogenesis. He was pessimistic about the prospects for a comprehensive theory of evolution, but dropped hints raising the specter of an unorthodox system of predetermined evolution. Hybridization was also accepted as a source of new species by a few Mendelians, a view pushed to its extreme by Lotsy. Although these ideas were not taken up by the majority of biologists, the fact that they were published and discussed reveals the depth of the confusion existing in evolution theory at the time and the role played by traditional values in promoting the general anti-Darwinian attitude of early geneticists.

The possibility that evolution might take place through a series of discontinuous steps, rather than through the gradual accumulation of minute variations, had been raised from the beginning of the Darwinian debate. T. H. Huxley wrote to Darwin after reading the *Origin*, complaining that "you have loaded yourself with unnecessary difficulty in adopting *Natura non facit saltum* so unreservedly."[1] In his review of the book, Huxley also insisted on the importance of saltations in natural evolution and their analogy to the artificial creation of breeds such as the Ancon sheep from a "sport of nature."[2] Throughout his career, he always believed that the problem posed by the discontinuity of the fossil record could be solved by supposing that the discontinuities were real steps in evolution. Much later, he even expressed approval of Bateson's support for discontinuous evolution, although it is doubtful that he would have accepted his rejection of the principle of utility.[3]

Darwin's cousin Francis Galton was also an advocate of discontinuous evolution. In *Hereditary Genius* (1869), Galton drew an analogy between evolution and a multifaceted stone, which might be rocked from side to side upon one of its faces until eventually it rolled over into a completely new position.[4] The analogy was formalized by means of a diagram of a geometrical polygon in *Natural Inheritance* (1889).[5] There Galton made it clear that although natural selection could act upon everyday individual differences (the wobbles of the polygon on a single face), it could not produce any permanent improvement of the race by this means because there was a tendency for each generation to regress to the mean value for the selected character. Only the large saltations (when the polygon moved onto a new face) were able to set up a new mean for the character, thus giving a quantum leap in evolution. Although Galton's work in the statistical analysis of variation paved the way for the biometrical school of Darwinism, his views on discontinuous evolution led him to side at least in part with the opponents of that school in the later debates over the mutation theory. Another curious point is that Galton's program of eugenics—the improvement of the human race by artificial selection—was founded upon a belief that such a process was not analogous to natural evolution and could thus only improve the tone of the existing race, rather than achieve any substantial biological progress.[6]

Galton's polygon analogy was taken up by St. George Jackson Mivart in *On the Genesis of Species*.[7] This raises an important question about the interpretation of the analogy, since almost certainly Mivart had a very different purpose in mind. His aim was to substantiate not just a theory of discontinuous evolution, but also one in which development was predetermined and hence teleological. Mivart stated that external factors produced ordinary variations (forces rocking the polygon from side to side) and eventually stimulated the production of a saltation (a bigger force pushing the polygon over onto a new face); but external factors were the stimulus and nothing more, since the direction of variation was internally predetermined (the new position depended upon the shape of the polygon). Thus, external factors had no control over the direction of evolution. It is clear, however, that both Huxley and Galton saw saltations as additions to the selection process, as new and more powerful sources of variation upon which selection for increased utility could act. They did not imagine the whole species suddenly adopting an entirely new form, irrespective of whether that form conferred adaptive benefits or not. Saltations were not produced in one direction, but were random in the sense required by Darwinism. Selection would determine which would succeed and which would be eliminated. Galton's original analogy of the three-dimensional stone was thus a better representation of his position than the two-dimensional polygon, since it implied that there were a number of adjacent faces onto which it could roll, not a single, predetermined future state.

The majority of investigators studying heredity and variation at the turn of the century were convinced that new characters appeared and were able to perpetuate themselves whether or not they had any adaptive value. They believed that it was indeed the internal laws of variation that directed evolution, even though these laws might not generate the absolutely predetermined sequences imagined by Mivart and the supporters of orthogenesis. If new varieties could appear and flourish in the garden or the laboratory, there was no reason why they should not do the same in nature. William Bateson, who linked saltative evolution with Mendelism, was a strong opponent of selection and the principle of utility. Hugo De Vries disagreed with the majority of his followers by supposing that selection might eliminate many mutations after they had been produced by internal forces. Field naturalists such as Edward B. Poulton lamented the emergence of a generation of biologists with so little understanding of the pressures under which animals and plants lived in the wild. To suppose that mimicry, for instance, was the product of coincidental variation in different species—as both Eimer and the Mendelians did—was typical of the way in which the hard lessons of Darwinian experience were being forgotten.[8] Yet the disrepute into which Darwinism had fallen through its failure to explain many other phenomena of variation ensured that the biologists who studied these phenomena would continue to reject the idea that adaptation was significant.

William Bateson began his career with a study in the classic morphological tradition, as adapted to evolution theory by Haeckel and his generation. He undertook a study of *Balanoglossus* with the aim of establishing the relationship of this primitive creature to the main divisions of the animal kingdom and thereby throwing light on the origin of the vertebrates. The culmination of this work came in 1886 with the paper "The Ancestry of the Chordata," the opening page of which recorded his growing dissatisfaction with this whole approach.[9] There Bateson expressed what would soon be the general feeling of a new generation of biologists: The creation of hypothetical genealogical trees, which, given the lack of fossils, were completely untestable, had become a totally sterile scientific exercise. Evolution theory needed to find a more fruitful approach, and Bateson now believed that the best way of doing this was to look more closely at the raw material of evolution; that is, variation. The study of *Balanoglossus* did, however, direct his attention to one kind of variation that might prove interesting. The main difference between this organism and a normal vertebrate was its lack of segmentation, and Bateson was left wondering whether the multiplication of parts by repetitive division might be responsible for such an evolutionary breakthrough.

At first, though, Bateson set out on a different path. In 1886–87 he went to the steppes of central Asia to study the correlation between environmental differences and specific variation. There he found a number of lakes with varying levels of salinity, which he saw as a means of testing the basic utilitarian postulate that the environment does determine the form of the species. Arthur Koestler, following Bateson's son, Gregory, interpreted his real purpose to be testing the Lamarckian hypothesis of the direct action of the environment upon the organism.[10] The one positive case that he reported on his return was indeed presented as an indication of the harmful effect of excess salinity upon a freshwater shellfish. The fact that Bateson wondered how long it would take the effect to disappear if the variant form were returned to fresh water suggests that he was thinking in terms of a hereditable nonadaptive variation stimulated by unfavorable conditions.[11] At the same time, however, he recorded that in many other species there was no sign of any correlation between variation and changed conditions. Bateson always regarded this trip as a failure because he found so few positive correlations, but in fact the negative results were completely to alter his attitude toward evolution. To Bateson, these findings did not just disprove Lamarckism; they undermined the basic utilitarian assumption of Darwinism as well. If the same species could live in very different conditions without showing any signs of adjusting to them, it was obvious that the environment did not exert the kind of control over organisms that the selection theory required. From this point on, as he became interested in the discontinuity of variation, Bateson was convinced that selection did not play a role in determining whether a new form could become established; it was the variation itself that was central in directing the course of evolution.

Whatever the Lamarckian purposes of his trip to Asia, there is evidence that Bateson was already a convinced hereditarian in his views on human affairs. A letter home in 1887 argued that heredity gave unequal qualities to each person, and that governments should take this into account.[12] Only a fear that the popular eugenics movement oversimplified science, coupled perhaps with the independent streak in his own character, prevented him from taking an important part in the efforts that were made during the early twentieth century to apply genetics to the improvement of the human race by selective breeding. Just the same, long before he became aware of Mendel's laws, Bateson had already begun to believe that variant characters were rigidly fixed by heredity.

In the meantime, Bateson began to concentrate on the nature of variation, becoming ever more impressed with the extent to which characters within a species varied in a discontinuous manner. Coupled with his hereditarian views, this would eventually leave him well prepared to appreciate the significance of Mendel's work. It is probable that his interest in discontinuous variation was stimulated by conversations with the American biologist William Keith Brooks, even before the Asian expedition. In *The Laws of Heredity* (1883), Brooks devoted a section to saltatory evolution in which he cited the views of Galton and Huxley on the question. Casting around for a new way of looking at variation after his failure to find a consistent link with the environment, Bateson became interested in an obvious case of discontinuous variation: the polymorphism of many species of flowers. By 1891, he was so impressed with the discontinuity of flower forms within single species that he found it impossible to accept the Darwinian claim that such different structures had gradually been produced by the divergent selection of minute variations. W. B. Provine has pointed out the similarity between the terminology used at that time by Bateson and that adopted by Galton to distinguish between individual differences and major saltations.[13] It was, however, more than a question of postulating a saltative origin for these characters: Bateson also displayed complete indifference to the Darwinians' efforts to postulate an adaptive purpose for the divergence of form.[14] In fact, the language in which he expressed his opposition to both the gradualism and the utilitarianism of the selection theory was already beginning to anticipate the style he would adopt in his major attack on Darwinism, *Materials for the Study of Variation* (1894).

In the preface and introductory chapter to this book, Bateson recorded his dissatisfaction with the morphological approach, the recapitulation theory, and unverifiable hypotheses on the ancestry and adaptive purpose of structures. While conceding that species must be largely adapted to their environment, he now stated his conviction that Darwinism would never succeed in explaining the details of each species' structure.

> We knew all along that species are *approximately* adapted to their circumstances; but the difficulty is that where the differences in adaptation seem to us to be approximate, the differences between the structures of species are frequently precise. In the early days of the Theory of Natural Selection it was hoped that with searching the direct utility of such small differences would be found, but time has been running now and the hope is unfulfilled.[15]

The same objection was also fatal to Lamarckism.[16] Bateson's own evidence suggested that most significant variation must be discontinuous. This would solve the problem created by the widespread existence of nonadaptive characters, provided one assumed that the environment had only limited ability to suppress useless forms produced by saltation. It would also solve the longstanding problem of how selection could produce adaptive characters when the incipient stages of those characters were too small to confer any advantage.[17] In other words, some new characters produced by saltation were immediately able to take on a function for which they turned out to be useful—although many others survived even though they had no use at all.

On several occasions, Bateson made the obvious point that the only way of proving the discontinuous origin of the variant forms was to find an offspring differing significantly from its parents.[18] If it was not possible to observe this relationship, it would be acceptable to cite examples where two or more distinct forms coexisted in the same species, and where there was no sign of intermediates by which one form could gradually have evolved from another. In fact, the hundreds of cases discussed by Bateson fell into this last category. His book was not a study of what De Vries would have called mutations, but of discontinuous variations within species (polymorphism in the more extreme cases), which were assumed to be the products of saltative evolution. Most of the examples were of what was called "meristic" variation, where the element of discontinutity was almost self-evident. In those characters such as the petals of a flower where there are a distinct number of complete units, the number of units can only vary discontinuously, by addition or subtraction. Here Bateson's interest in the repetition of identical parts, first sparked by the lack of segmentation in *Balanoglossus,* reemerged as a powerful factor directing his opinion of what was the most important kind of variation for evolution theory.

Materials deliberately avoided comment on two interrelated issues that Bateson felt it would premature to tackle at that stage, the mechanisms whereby the saltations were produced and inherited. On the topic of inheritance, however, he did say that "we shall not go far wrong in assuming the possibility that [such variations] may reappear in the offspring."[19] Indeed, to avoid the argument that Fleeming Jenkin had once raised against Darwin, saltations would have to be inherited as units if they were to become established in the species. In the late 1890s Bateson began a series of experiments on the hybridization of varieties with distinct character differences, and he was thus well prepared to appreciate the significance of Mendel's paper when De Vries and others drew attention to it in 1900. He became a leading champion of Mendelism and began to gain a wider reputation as the new theory of heredity caught on. *Materials* had not sold well, perhaps because the anti-Darwinian movement was slow to get off the ground in Britain, but it had already alienated Bateson from a former friend, W.F.R. Weldon, and from the whole biometrical school.[20] Bateson's support for discontinuous, nonadaptive evolution had led to bitter controversies with the Darwinists in the late 1890s, and Mendelism became yet another source of dispute. Bateson naturally identified the unit inheritance of discontinuous characters as evidence that saltations could immediately establish themselves in a species to produce distinct evolutionary steps. The biometricians, at least in part because of the intensity of the feelings aroused by the earlier debates, could not appreciate that the recombination of a number of Mendelian factors affecting a single character would generate the continuous ranges of variation that they were studying. While Bateson was hailing Mendelism as the key to a new understanding of heredity and evolution, his opponents continued to insist that discontinuous variation was an anomaly of no real significance.

De Vries soon lost interest in Mendelism and began to insist that his mutation theory explained the origin of new varieties and species. Bateson did not believe that a single mutation could create a cluster of new characters sufficient to define an entirely new species or subspecies. Indeed, he had grave doubts as to whether mutation could do more than destroy an existing Mendelian factor. His somewhat confused ideas on the origin of saltations were given in the Silliman Lectures at Yale University in 1907 and were subsequently worked up into the *Problems of Genetics* in 1913. Here Bateson repeated all his old anti-Darwinian arguments; he gave more examples of discontinuous variation, which he did not think could be explained in terms of selection and geographical isolation. He was not optimistic about the prospect of developing a complete theory of the origin of variation, claiming that although no one doubted evolution, "few who are familiar with the facts that genetic research have revealed are now inclined to speculate as to the manner in which the process has been accomplished."[21] Yet he did offer a few hints on future developments, some of which anticipated the unorthodox developmental genetics of Richard Goldschmidt.[22] He held that discontinuous meristic variation disproved the theory according to which each Mendelian factor was attributable to a chemical unit in the germ plasm (the gene, as Johannsen had first called it in 1909).[23] Instead, he offered an entirely different view of the nature of inheritance and variation, one based on the concept of rhythmical forces controlling cell division.

Bateson's opposition to the chromosome theory of heredity led to his estrangement from the main path of genetic research pioneered by Thomas Hunt Morgan and others. In many respects, his attitudes remained fixed in the nineteenth-century mold, which prevented him from seeing the direct equation of a hereditary factor with a chromosomal unit as anything more than simpleminded materialism. His own epistemology was of the intuitive variety, and it has been suggested that he may have been influenced by the idealist philosophy of the physicists who dominated Cambridge science during his years there.[24] The phenomenon of meristic variation convinced him that chemical units in the germ plasm determined only part of the transmission of heredity. More crucial was the process of cell division, which Bateson believed to be under the control of unknown physical forces imposed upon the material structure of the fertilized ovum. Such forces were vibrational in character and gave rise to meristic phenomena, striped coloration, and so on, by exactly the same process as waves making ripples on a beach.[25] A reference to the Chladni figures produced in sand upon a vibrating plate made the same point, and they also allowed

Bateson to speculate that the nature of this vibrational factor was such that it permitted only discontinuous variation from one stable pattern to another.[26]

Bateson's view that physical forces affecting the growth process might have a role to play in variation seems almost to parallel D'Arcy Thompson's approach to morphology. Certainly it brings out the apparently paradoxical fact that a leading experimentalist could still identify with a conservative, almost idealist viewpoint more characteristic of orthogenesis. The question, however, remains: Were the discontinuous changes in the pattern of development stimulated by internal or external factors? There would obviously have to be an internal component, since the existing pattern of vibration would define the alternative modes into which it could be switched. Note that the analogy with the Chladni figures, by exploiting the idea that the system was predisposed to vibrate in a certain number of modes, emphasized the discontinuity of variation more than orthogenetic regularities. Yet, just as in the case of the changing vibrations that excited the figures, there must have been an external stimulus that elicited the variation. Bateson saw that the experimental production of monstrosities by altering the environment in which the embryo developed threw light on this problem. Presumably, an altered environment suppressed or elicited a different mode of vibration in cell division, as, for example, in C. R. Stockard's experiment in which a deficiency of magnesium caused the growth of one-eyed fishes.[27]

By now Bateson was totally opposed to the Lamarckian position that the adaptive response of the adult organism to its environment could be inherited; hence, his bitter criticism of Kammerer.[28] Nevertheless, he was prepared to allow the environment a role in stimulating nonadaptive patterns of growth. He cited numerous examples of this effect, including the widely discussed experiments of W. T. Tower on the potato beetle.[29] He criticized Semon for missing the point that most of the "acquired characters" he cited as evidence of Lamarckism were in fact nonadaptive.[30] As we have seen, some of the nonadaptive characters that Bateson attributed to external influence were to be explained as modifications of the growth forces, and it is not clear how he could have thought these might become hereditable. There were, however, other characters that he did attribute to the chemical nature of the germ plasm although, unlike the supporters of the chromosome theory, he thought they were controlled by the overall arrangement of the particles. Such characters could also be modified by external factors acting directly on the germ plasm, although in this case Bateson saw the environment as a destructive influence that poisoned the reproductive cells and produced stunted or deformed offspring. He quoted Virchow's maxim that all variation is in origin pathological and cited the experiments of those who showed the inheritance of degenerative characters produced by alcohol poisoning and other means.[31] The fact that Bateson, a leading opponent of Lamarckism, should have freely accepted these experiments shows how desperate the Lamarckians must have been to cite the same evidence on their own behalf.

The concept of destructive variation brings us to another strange product of Bateson's belief that genetics had closed off most of the traditional explanations of how evolution worked. In his presidential address to the Australian meeting of the British Association in 1914, he hinted at a theory that would reduce all evolution to the unfolding of a predetermined set of characters. Bateson thought that there was no such thing as a positive mutation in the germ plasm, that is, a mutation creating a new character. Changes in the germ plasm were always destructive because they led to the elimination of a previously existing Mendelian factor.[32] To explain why new characters did appear from time to time, he postulated the widespread existence of inhibiting genes, which blocked the expression of other characters. When an inhibitor was destroyed by negative mutation, the character it had once masked would appear in the species. This at least raised the possibility that the whole of evolutionary "progress" might be the effect of a genetic degeneration, which allowed the gradual appearance of characters that had all been present in the original forms of life, along with the appropriate inhibitors.

> If then we have to dispense, as seems likely, with any additions from without, we must begin seriously to consider whether the course of Evolution can at all reasonably be represented as an unpacking of an original complex which contained within itself the whole range of diversity which living things present. . . . At first it may seem rank absurdity to suppose that the primordial form or forms of protoplasm could have contained complexity enough to produce the divers types of life. But is it any easier to imagine that these powers could have been conveyed by external additions?[33]

A little later Bateson suggested that the order of unpacking might be predetermined, which would cause the regular development of characters that Eimer had called orthogenesis.[34] This served only to strengthen the implication that the entire evolutionary process had been nothing but the unfolding of a preexisting pattern (which is, in fact, the original meaning of the term "evolution," now forgotten). However puzzled Bateson's contemporaries might have been about the true nature of the process, they were not about to follow his retreat into this position, which some compared to the old preformation theory of embryology.[35] Bateson's desperation because of the refusal of his own interpretation of genetics to yield anything resembling a mechanism for progressive evolution had led him into a position that resembled little more than a throwback to the old vision of a divine plan built into the very nature of life. Meanwhile, the majority of geneticists had accepted the concept of positive mutations and were building it into their theories of evolution.

Before turning to the mutation theory, however, it is necessary to look briefly at another unorthodox extension of Mendelism into the field of evolution theory. In discussing De Vries's evidence for mutations in the evening primrose, Bateson (quite correctly, as it turned out) suggested that this species had an abnormal genetic constitution and might be a hybrid.[36] The alleged mutations were thus not genuinely new characters at the genetic level, but merely hitherto unnoticed combinations of factors brought together by the normal process of reproduction in the hybrid. Other Mendelians adopted a similar viewpoint, but it was the Dutch botanist J. P. Lotsy who in 1906 elevated hybridization to the status of a major cause in evolution. Like Bateson, Lotsy did not agree that new genetic factors could be created; but he had a clear appreciation of developments that had been taking place in systematics, particularly Karl Jordan's recognition that each Linnaean species was composed of many true-breeding subspecies, or "Jordanons." The Linnaean species, he held, was a complex of characters originally derived from the hybridization of two distinct forms. Mendelian segregation gradually broke the complex down into its constituent parts, the Jordanons, many of which would eventually be eliminated by natural selection. With apologies to Bateson, Lotsy argued that Mendelism referred not to the inheritance of a character, but to its disintegration—yet it was Mendel who had provided the clue to the origin of new species by showing how two dissimilar forms could be hybridized to bring forth a number of gametes of different constitution.[37]

Modern biologists believe that new species may occasionally arise by hybridization, but for Lotsy this was the only process needed to explain not just the origin of species, but also the origin of classes and phyla. He held that the vertebrates had first appeared through the hybridization of two invertebrate types, while the reptiles and mammals had similarly been created by crossing within the early vertebrates. The fossil record was consistent with this view because it showed (1) that the origin of the great classes lay far back in time, (2) that each appeared suddenly with a great diversity of forms, and (3) that many of these early forms were gradually eliminated in the course of time.[38]

Lotsy realized that this would make nonsense out of all previous attempts to reconstruct the course of evolutionary progress, including his own work in plant phylogeny, but he insisted the possibility should not be ruled out simply because it required so drastic a revision of previous interpretations. He was even prepared to suggest that the human race might have arisen directly from the hybridization of two *reptile* species, thus bypassing the rest of the mammals altogether![39] The idea of evolutionary progress was thus a misconception: There might have been a predetermined order in which the successful crosses had to occur, but this did not lay down a really progressive sequence. The most important evolutionary breakthroughs had all occurred long ago, while later hybridizations within the major classes produced only more limited changes, and now seem to have stopped altogether. Lotsy speculated that he might, in fact, be observing "the last trial of life to maintain itself on our globe," after all its potentiality for creative hybridization had been exhausted.[40]

Like Bateson's theory, Lotsy's implied that the pattern for the history of life was originally contained in the genetic potential of a few forms. Whether the pattern was revealed by new combinations of existing elements or by the elimination of inhibiting factors, the element of preformation was still the same. Significantly, Bateson did not conceal his sympathy with Lotsy's approach.[41] Not surprisingly, they did not gain much support for their unorthodox applications of Mendelism to evolution. Although most Mendelians were just as dissatisfied with conventional evolution theories, they could not go along with this wholesale rejection of the basic philosophy of progress. Bateson and Lotsy illustrate the extremes to which a new approach can be pushed when, in its early stages, it seems to attack the structure of all existing knowledge. The really surprising thing is that there were so few outright condemnations of their ideas as a threat to the scientific investigation of the history of life. This indicates that others shared their confusions, even though they sought a solution to the problem of evolution through the more straightforward concept of progressive mutations.

DE VRIES AND MUTATIONS

Modern geneticists think of mutations as alterations in the structure of single genes, which introduce new characters into the breeding population and hence increase its variability. The genetical theory of natural selection assumes that only those mutations causing small character changes will have the chance of being favored by selection so that their frequency is increased within the population and they take part in the evolution of the species. This, however, was not how the term "mutation" was used in the first decade of the twentieth century. Hugo De Vries first popularized the term by using it to denote large-scale genetic changes capable of producing a new subspecies, or even species, instantaneously. Selection of small individual differences thus had no importance in evolution, although De Vries himself believed that selection would still allow only those mutations conferring some adaptive advantage to flourish in the long run. Many of his followers, however, adopted Bateson's view that the demands of the environment were not that rigid, so that mutations that did not produce any advantage could direct the course of evolution.

From De Vries's original theory we can trace two important ways in which biologists' attitudes toward mutations changed. First, the size of the mutations that were supposed to affect evolution was gradually reduced. Instead of being wholesale genetic renewals on a scale that established a new breeding population overnight, the mutations came to be seen as changes in single genes that had only a limited effect on the form of the organism and fed into the existing breeding population. Second, since it was no longer possible to see the mutated form as breeding apart from the parent, it was gradually appreciated that selection might affect the success or failure of a mutated gene, either spreading it into the population or eliminating it. From this last point developed the genetical selection theory, although many early geneticists did not give selection a major role, and even those who did accept it tended to adopt a rather simplistic view of its operation.

The question of the size of the mutations would at first sight appear to be a straightforward one. De Vries overestimated their size and frequency because his evidence was based on an anomalous subject, the evening primrose, *Oenothera lamarckiana*. In fact, what he was observing were not true mutations, but the products of genetic recombination in a very unusual hybrid species. Although biologists eventually realized that there was no chance of finding "mutations" on this scale in other species, observation revealed instead the presence of smaller-scale true genetic mutations, especially in the fruit fly, *Drosophila*. A connection between the sudden appearance of new characters and the Mendelian scheme for the unit inheritance of existing characters was soon perceived, one that allowed the emergence of the modern concept of a mutation as an alteration in the structure of an existing gene. At first, however, there was still a tendency for the Mendelians to exaggerate the size of the mutations that might affect evolution. Only with the work of R. A. Fisher and others in the 1920s did it become apparent that selection could act on very small character differences, and that most large mutations were lethal or detrimental in any case. The majority of field naturalists and paleontologists remained convinced of the gradualistic nature of evolution and were thus hostile to the work of the geneticists until this scaling down of mutations had taken place. Curiously, though, the few naturalists who did take De Vries's idea seriously tended to go off in exactly the opposite direction. Because their chief concern was the fundamental question of the origin of phyla and classes, they postulated vast mutations on a scale far beyond anything that even De Vries had imagined.

The question of whether mutations of any size are subject to selection is complex, since it touches on the most sensitive implications of Darwinism. Many mutationists shared Bateson's hostility to the rigid utilitarianism of the selection theory and saw their alternative as offering a way out of the sterile attempts to assign adaptive values to every character. De Vries agreed, up to a point, but held that it was essential to allow the environment a part in deciding the eventual survival of a mutated form. To abandon this requirement was to risk losing the most crucial methodological advance made by Darwin: his rejection of teleology. De Vries shifted the level of the struggle for existence from individual competition to a struggle between well-established varieties or species. His theory was thus still of value in esscaping some of the most pressing moral and philosophical problems raised by Darwinism. It no longer gave support to the laissez-faire, individualistic form of social Darwinism, while his followers eliminated the need for struggle at any level. Loren Eiseley suggested that the mutation theory flourished at least in part because it avoided the more repugnant aspects of Darwinism that were now perceived as reflections of sordid Victorian materialism.[42] The young Thomas Hunt Morgan repudiated Darwinism altogether in an effort to escape from the image of nature progressing through struggle, and he was later converted to selectionism only in a form that did not imply the constant elimination of the unfit. The great advantage of the mutation theory was that unlike orthogenesis, it did not replace the struggle for existence with an internal force so rigidly predetermined that it drove the species to extinction.

In 1889 Hugo De Vries published a theory of "intracellular pangenesis," a highly modified version of Darwin's views on heredity. The theory led him to the idea of unit characters and thus paved the way for a series of breeding experiments and the rediscovery of Mendel's laws in 1900.[43] In addition, De Vries was interested in the origin of new characters. As early as 1886 he had noticed *Oenothera* growing wild in several different forms. When cultivated, the plant appeared to throw off true-breeding mutations, which differed enough from the parent form for De Vries to give them their own specific names. The theory that evolution arose from such mutations was published in *Die Mutationstheorie* (1901-3). In 1904 he gave a series of lectures at the University of California, which subsequently appeared under the title *Species and Varieties: Their Origin by Mutation.*[44]

De Vries said he was working within the Darwinian tradition: "My work claims to be in full accordance with the principles laid down by Darwin, and to give a thorough and sharp analysis of some of the ideas of variability, inheritance, selection and mutation, which were necessarily vague at his time."[45] He interpreted Darwin's own opinions in a way that allowed him to draw a parallel between them and his new theory. According to De Vries, Darwin himself had originally wanted to give saltations a role in evolution, but had been dissuaded from this by the argument that such individuals would be swamped by the mass of the unchanged population when they began to breed.[46] This argument had seemed decisive because Darwin was thinking of single individuals with the mutated character, but De Vries was convinced that a mutation appeared in many individuals at the same time.[47] A new variety or subspecies was thus formed all at once and continued to breed true within its own population. Darwin had believed that varieties were formed gradually by the selection of minute individual differences, but De Vries insisted that this process could produce only local races that would revert to the original type as soon as they were removed from their characteristic environment.[48]

Since true-breeding varieties were not formed by selection, their appearance could not be dictated by utilitarian considerations. It was possible for a new mutation to appear that gave no advantage—yet as long as its character was not actually harmful, it would be able to establish itself because it was present in large numbers. De Vries admitted that "useless, but not dangerous mutations must appear as often as useful ones, and have almost as much likelihood as these of persisting."[49] These useless characters, which the taxonomist often exploits as a means of distinguishing closely related species, were the result of retrogressive mutations, while progressive mutations were responsible for the more positive developments in evolution.[50] De Vries also believed that species exhibit periods of rapid mutation alternating with periods of stability—*Oenothera* was useful because it was at present experiencing an active phase.[51] During such periods evolution could proceed quite rapidly, and the entire history of life could have unfolded in a much shorter period of time than Darwin had supposed, thus checkmating Lord Kelvin's argument based on the age of the earth.

De Vries's claim to be a good Darwinian was based on the fact that his mutations were supposed to form a new starting point for natural selection. In Darwin's theory, selection acted upon random individual differences, while in De Vries's it worked among the mutations. During its active phase, a species would split into a number of distinct sub-species, all occupying the same area, and these different forms would have to compete with one another to determine which would replace the parent species. Although nothing could prevent a mutation with the same adaptive value as the parent form from becoming established for a while, this was only a temporary state of affairs. Sooner or later a progressive mutation of real value would appear and would use its advantage to displace both the parent form and all the retrogressive mutations that had appeared up to that time. According to De Vries, "By far the greater number of mutations presumably perish, nipped in the bud by natural selection. Other forms may continue for one or two years, but after a time they too disappear. It is only a very few which ultimately come to take part in the great struggle for existence."[52] Mutation normally formed a series of distinct subspecies (although De Vries did give the new forms of *Oenothera* specific names of their own). Truly separate Linnaean species that could be recognized as such by the taxonomist would be created when enough intermediate mutations had been eliminated by selection.[53] This process would probably occur so rapidly that the details would not be preserved in the fossil record— hence its apparent discontinuity.

De Vries believed that his retention of selection at this level was of vital importance in preserving the scientific credibility of the theory. He pointed out that the key to Darwin's elimination of the old argument from design was his insistence on the random nature of the variations upon which selection acts. Adaptation did not prove divine benevolence because there was nothing to suggest that favorable variations appear more readily than unfavorable ones. "Every hypothesis which differs from Darwin's in this respect," he wrote, "must be rejected as teleological and unscientific."[54] The mutation theory was safe in this respect since it "demands that organisms should exhibit mutability in almost all directions."[55] Some of De Vries's followers seem to have thought that he favored the idea of mutations consistently aimed in particular directions, but in fact he explicitly attacked W. B. Scott's paleontological argument for Waagen's directed mutations.[56] De Vries insisted that orthogenesis was an illusion created by the action of selection, which rapidly suppressed all mutations except those that happened to lie in the direction of increased specialization.

One American supporter of the mutation theory claimed that it allowed us to see how Adam and Eve gained their distinctive human characteristics.[57] There is no evidence to suggest that De Vries himself had any desire to solve the theological problem of the status of the human species in this way, but he did realize that his theory undermined the analogy upon which social Darwinism was based. The variations between individual human beings were mere fluctuations with no evolutionary significance, so there would be little point in allowing a free-for-all struggle to weed out the weakest. Darwin's theory, he argued (meaning his own version of Darwinism), "has no bearing whatsoever upon our present socio-anthropological inquiry."[58] De Vries even maintained that the human races were only local varieties, not the products of distinct mutations, thus destroying the scientific argument for racial inequalities.[59] The unpleasant social implications of Darwinism were thereby avoided, not by eliminating struggle from nature, but by shifting it to a level at which no analogy could be drawn to human affairs.

Many of the American supporters of the mutation theory seem to have regarded it as a complete alternative to Darwinism. This is particularly the case with the young Thomas Hunt Morgan, whose *Evolution and Adaptation* (1903) devoted as much space to attacking the principles of Darwinism as it did to promoting the mutation theory.[60] At this point in his career, Morgan had not accepted Mendelism, and he was thus free to adopt De Vries's interpretation of mutations as instantly created subspecies. Nevertheless, he would not agree that there was any need to retain the struggle for existence among the mutated forms. Natural selection was a mechanism for explaining the origin of adaptations, not the origin of species, and Morgan was determined to show that adaptation was of only minor significance in evolution. Like Bateson, he agreed that every form had to be more or less adapted to its environment, but he denied that the level of adaptation was consistent with the claim that natural selection had guided the course of evolution. Morgan believed that the pressure of selection was so slight that nonadaptive mutations would face no threat of extinction unless they were actively harmful. The mutations themselves would thus direct the course of evolution. What had been for De Vries the relatively trivial ability of mutations to produce nonadaptive characters on a temporary basis was now seen as the foundation of a complete alternative to selection.

Morgan argued that the Darwinians simply assumed that every

character had a function, even though no evidence could be brought forward to back up their assertion. Why, he asked, should some animals appear to be camouflaged, while others are brightly colored? Perhaps the camouflage is not what it seems. In the case of microscopic organisms, color is clearly of no significance, so one should be suspicious of the claim that it is always of some value to larger creatures.[61] How can the patterns on the inside of a snail's shell be of any use (an example also used by Eimer)?[62] The same problem arises in the case of dimorphism: Even Darwin could not explain why *Primula veris,* for instance, exists in two different forms (this kind of example may have come from Bateson).[63] Can adaptation explain the almost universal occurrence of some kind of symmetry in animals?[64] In other cases the argument for utility proves too much and reduces itself to absurdity. It might be of some use for a butterfly to disguise its wings as a dry leaf, for instance, but what is the value of adding extra details such as an imitation midrib?

The purpose of these arguments is plain: If Morgan could show useless characters to be as common as adaptive ones, then selection would no longer provide a viable explanation of evolution. The fact that selection and not utility was Morgan's real target is confirmed by his efforts to undermine the credibility of the Darwinian struggle for existence. It would be necessary, he claimed, for the Darwinians to prove that variation really could have an effect on the death rates of the individuals concerned. This is exactly what the biometricians were trying to do, but Morgan believed that there was an indiscriminate destruction of the young in each generation long before adult characters could have any effect.[65] He claimed that through the elimination of less well-adapted forms, selection must always lead to the increased specialization of any character; yet often we see a range of forms with different levels of development in the same character. Thus some ants have soldiers, which are supposed to be of use in defending the nest, but even if we admit that they do serve this purpose (which Morgan doubted), we cannot explain why related species manage to exist without soldiers.

> Neither need we suppose that during the evolution of these colonial forms there has been a death struggle accompanying each stage in the evolution. If the members of a colonial group began to give rise to different forms through mutation, and if it happened that some of the combinations formed in this way were capable of living together, and perpetuating the group, this is all that is required for such a condition to persist.[66]

A similar argument could be applied to secondary sexual characters. Morgan devoted a chapter of *Evolution and Adaptation* to the claim that competition among males to obtain females could not produce these characters, which must therefore be the result of mutation. Without sexual selection, Darwin's system crumbled, since by itself natural selection could not explain the origin of sexual characters that were of no use to the species.

Morgan's position was based on the claim that many organs have no adaptive value, but without the theory of selection he was left with the problem of accounting for those that are in fact adaptive. He did not want to argue that a well-developed organ could be produced by a single mutation, so it became necessary to postulate a mechanism that would guide mutations in a particular direction—in effect, a mechanism of orthogenesis. Morgan speculated that the internal constitution of the organism might predispose it to mutate only in certain ways. He acknowledged that there was as yet no experimental evidence for this, but insisted that the possibility was worth further study. One would not, of course, want to adopt anything as mystical as Nägeli's perfecting principle, but the mutation theory could not afford to ignore the implication that something within the organism was responsible for producing evolution.[67] The process would be random in the sense that the trends might lead in many directions, according to the constitution of the particular organism. Morgan had no interest in vast unifying trends on the scale postulated by the supporters of true orthogenesis. Selection might play the purely negative role of eliminating those trends that led in a harmful direction, but the trends themselves would control the production of both useful and adaptively neutral organs.

De Vries's claim that selection was still needed to explain the adaptive trends was thus superfluous. The crucial distinction between the two interpretations of the mutation theory was that Morgan refused to admit the constructive power of differential survival rates at any stage in the evolutionary process. Anything that could survive, however marginally, could do so—there was no tendency for the better adapted to drive out the lesser, and hence no real struggle in nature. The mutation theory made it possible to escape from the "dreadful calamity of nature, pictured as the battle for existence."[68] In these words Morgan revealed his distaste for the Darwinian vision of progress through struggle. At this point, he drew no explicit connections with the social implications of such a philosophy; instead, his objections seemed to lie at a more general moral level. In the end, all of his arguments against utility served the purpose of eliminating the need to see nature itself as founded on struggle and suffering.

As late as 1910 Morgan repeated his suggestion that directed evolution resulting from a series of internally controlled mutations might create adaptive structures without selection. The title of that paper, "Chance or Purpose in the Origin and Evolution of Adaptations," illustrates his desire to promote a more orderly image of the process than was possible with the selection theory. Scientists frequently made the general point that the mutation theory replaced chance with law;[69] but for Morgan, an internal cause of evolution was preferable only as long as it was an ordering factor that generated something more than undirected mutations. This side of his argument was echoed by Maynard Metcalf, a prominent exponent of the claim that orthogenesis was needed to supplement Darwinism in order to eliminate the problem of the incipient stages of adaptive structures.[70] A link between orthogenesis and mutations was also drawn in R. Ruggles Gates's 1915 exposition of De Vries's theory. Gates was still convinced that the mutated forms of *Oenothera* were genuinely new species produced by changes in the chromosomes. Noting that Morgan had by this time demonstrated the occurrence of much smaller mutations that fed into the breeding population of *Drosophila,* he suggested that these were due to chemical rearrangements of particular genes, while De Vries's much larger mutations involved the change of an entire chromosome.[71] Gates discussed at length the evidence for orthogenesis and implied that De Vries himself believed that mutations would occur more readily in some directions than in others.[72] Since some orthogenetic trends seemed to go beyond the point of adaptation, selection could not be the directing factor—one had to look instead for an internal cause that would presumably derive from the chemical structure of the genetic material.

This interest in directed mutations suggests that the theory drew some of its support from biologists concerned about the issues surrounding orthogenesis. Both approaches stressed the internal causation of evolution, and although there was little hard evidence for directed mutations, there was a constant temptation to explore this line of thought in the hope of creating a more orderly picture of evolution than was possible with De Vries's original idea of purely random mutations. The work on *Drosophila* that was begun by Morgan and his students in the "fly room" of Columbia University in 1910 was thus important to evolution theory in two ways. Most obviously, it confirmed that mutations occurred in single genes, introducing new characters into an existing species rather than creating a new species altogether.

De Vries's original theory soon crumbled under this wave of hard evidence for much smaller genetic changes and the growing suspicion that what had been thought to be mutations in *Oenothera* were in fact the products of hybridization. Although Gates still accepted *Oenothera* as a mutating species in 1915, he was forced to record increasing dissatisfaction with this interpretation. Bateson had first hinted that *Oenothera* was a hybrid in 1902 and had repeated this claim in 1913 in his *Problems of Genetics*. By 1910 Bradley Davis had already begun a series of investigations supporting this interpretation.[73] Although the true genetic nature of *Oenothera* would not be unraveled until the work of R. E. Clelland in the 1920s, De Vries's interpretation had generally been abandoned long before this.[74]

Equally significant in the long run, however, was the implication of Morgan's work on *Drosophila* for the theory of directed mutations. By the time he wrote his *Critique of the Theory of Evolution* in 1916, Morgan was already seeking to explain adaptive trends as by-products of a simplified form of selection.[75] His work on genetic mutations was presented as evidence for the occurrence of random variation in many characters, which would have to be acted upon by selection for evolution to take place. Thus the laboratory evidence rapidly destroyed what had at first seemed to be an interesting combination of Waagen's and De Vries's definitions of mutation. Once the random nature of mutation became inescapable, it was only a matter of time before De Vries's original emphasis on a role for selection would have to be revived as a means of imposing a direction upon evolution. In the end, few biologists were prepared to contemplate a vision of evolution as a series of random (that is, unrelated) mutations. It may well be that the outburst of opposition to selection in the first decade of the new century was made possible only by the hope that a fusion of the mutation theory and orthogenesis might create the basis for a picture of orderly evolution. Once that hope had been frustrated by the evidence for the random nature of true mutations, a return to selectionism was inevitable after the initial hostilities generated in the debate with biometry had cooled off.

Before looking at the reemergence of the selection theory, however, it is important to note the way in which a few field naturalists and paleontologists made use of De Vries's original concept of mutation. In general, the wave of enthusiasm for saltative evolution was ignored by a generation of naturalists who had grown up thinking in terms of geographical isolation as an important factor in speciation. As C. Hart Merriam complained in 1905, a few examples of sudden mutation did not mean that all the indirect evidence for gradual evolution should be set aside.[76] Merriam referred to David Starr Jordan's emphasis on geographical isolation as the crucial factor that allowed the build-up of differentiating characters—although it may be noted that Jordan did concede a place for nonadaptive mutations in producing these characters.[77] Joel Allen, a prominent neo-Lamarckian, also endorsed the mutation theory as an explanation of geographical variation, thereby confirming the extent to which the American school had stressed both discontinuous steps and nonadaptive factors in evolution.[78] The widespread belief that some characters are nonadaptive was perhaps the only reason why some field naturalists might have toyed with De Vries's concept of mutation. Many, however, continued to see such characters as the product of gradual modification through the inheritance of nonadaptive changes in the organism when exposed to new conditions.

The most extravagant form of saltation was proposed by those naturalists concerned with the problem of the origin of phyla and classes. For them, conventional gradualistic theories seemed at their weakest; there was little fossil evidence for the gradual evolution of completely new types, and it was difficult to see how every step in the appearance of a new form could be of some adaptive value. Perhaps De Vries's mutations were only a small-scale example of much larger saltations that had occasionally appeared in the history of life to spark new directions in evolution. Unfortunately, there was no evidence of mutations on this scale occurring at any time during the more recent geological epochs. This inevitably led to speculations that the level of mutational activity had gradually died down in the course of geological time. Because such ideas were difficult to test, they raised once again the specter of mysterious forces operating only in the past, forever beyond the reach of laboratory study. They also ran counter to the trend in genetics, in which research was confirming the regular occurrence of mutations within each species. In effect, these exaggerated concepts of mutation bore no relation whatsoever to the increasing precision with which the term was being defined in genetics. Naturalists attempting to revive the old tradition of evolution by major saltations were able to borrow the term and thus gain a ride on the coattails of the new theory because De Vries's early misconceptions could be extended in a way that the laboratory biologists were now refusing to sanction.

Most paleontologists worked with evolutionary developments occurring within particular families or orders and were by now convinced that this kind of evolution was essentially a gradual process. Enough new discoveries had been made since the time of Darwin to suggest that the remaining "gaps" in the fossil record were illusions created by lack of information. Many believed that the same point held for the more funadamental gaps that marked the origin of new phyla and classes, but the comparative lack of hard evidence to support the evolutionary bridges required by this interpretation was beginning to disturb some paleontologists. There were a few key fossils, of course, especially *Archaeopteryx,* which pointed to an evolutionary link between the birds and the reptiles. Nevertheless, this important contribution to the original Darwinian debate had not been followed up, and by the end of the century some paleontologists were beginning to wonder if the sudden appearance of new forms in the record might be a genuine indication of major saltations. It was exactly this lack of confirming evidence that soured Bateson's attitude toward his morphological speculations on the origin of the vertebrates. Where a morphologist, however, could simply abandon the attempt to reconstruct evolutionary history and look for a more realistic project to tackle, a paleontologist had no choice but to confront the problem of interpreting the gaps in the record.

German paleontologists took a lead in exposing the lack of progress in the search for fossil evidence linking the classes. The most outspoken was Karl von Zittel, who created some excitement at the 1894 International Congress of Geology by questioning the assumptions made by the evolutionists.[79] He doubted that *Archaeopteryx* was a significant link between the birds and reptiles, and he pointed to the lack of any parallel evidence for similar links elsewhere. Although sympathetic to the general idea of evolution, he cautioned that at this level it still had to be treated very much as an unverified hypothesis. From this it was only a short step to an exploration of the possibility that saltations might account for the sudden appearance of new types and classes. De Vries's work seems to have stimulated some paleontologists to think along these lines in the early twentieth century. Most important was O. H. Schindewolf, whose *Paläontologie und Genetik* (1936), a late example of the trend, tried to synthesize the two sciences around the concept of large mutations. Schindewolf argued that single genetic mutations affecting the early stages of ontogeny might result in large changes in the adult form because the plasticity of the growth process allowed the organism to adjust to the major reorganization of its structure necessitated by the change. He openly stated that the first bird was hatched from a reptile's egg.[80] The fossil evidence, he believed, suggested that new characters did indeed appear first in juveniles and only later in adult forms—an exact reversal of the acceleration of growth required by Lamarckism.

At first it had not seemed too outrageous to suppose that quite small germinal changes might bring about major evolutionary steps. In a 1905 discussion of the mutation theory given before the American Society of Naturalists, the cytologist Edwin G. Conklin had also suggested this as a mechanism for the origin of new phyla.[81] Paleontologists such as Schindewolf were thus attempting to fuse what at first had seemed a plausible extension of the mutation theory with the fossil evidence for major discontinuities in evolution. By the 1930s, however, the majority of geneticists had decided that large-scale mutations could not affect evolution because they were invariably lethal. Instead, geneticists opted for a renewal of gradualism by postulating that natural selection acted to increase the frequency of new genes that conferred some slight advantage in reproduction.

The only prominent geneticist to retain an interest in saltative evolution at this time was Richard Goldschmidt. He remained convinced that the selection of minute genetic differences could only create subspecies (microevolution), but could not achieve the production of genuine new species (macroevolution). Like Schindewolf, he believed that a mutation affecting the early stages of growth would be able to produce a substantially new form by switching ontogeny into a new pattern of development. Goldschmidt conceded that such mutations were normally lethal because the growth process was so finely balanced that any significant disturbance led to a complete breakdown. At the same time, he insisted that occasionally a new pattern of growth, viable all the way through to maturity, produced a "hopeful monster" as the founder of a new species. In exceptional circumstances, such a drastic change might open up a whole new line of evolution; but in general Goldschmidt insisted on macromutations as the source of every new species, even within existing groups. His interest lay not so much in the origin of new types as in the ability of the developmental process to control existing lines of evolution. Once a viable pattern of growth was established, it would impose strict limits on what kind of new mutation was compatible with it, and might thus direct evolution along orthogenetic lines.

> The selection of the direction in which genetic change may push the organism is therefore not left to the action of the environment upon the organism, but is controlled by the surroundings of the primordium in ontogeny, by the possibility of changing one ontogenetic process without destroying the whole fabric of development. . . . Thus what is called in a general way the mechanics of development will decide the direction of possible evolutionary changes. In many cases there will be only one direction. This is

orthogenesis without Lamarckism, without mysticism, without selection of adult conditions.[82]

In recent years a few biologists have once again begun to advocate an epigenetic view of evolution derived from the ideas of Schindewolf and Goldschmidt on macromutations.

Most naturalists studying the geographical distribution of species were strongly in favor of gradualistic evolution under the influence of isolation, but even here there were a few critics who believed they had evidence that major saltations had originated new forms. H. B. Guppy was a British botanist who specialized in the distribution of plants on oceanic islands. On the strength of work performed in the Pacific during the 1890s, he advanced the theory that plants had been forced to adapt to a gradual drying up of the world's climate. Species and genera were merely eddies in a stream, the course of which was fixed by more fundamental biological processes.[83] In 1917 Guppy extended his world-dessication theory into a complete anti-Darwinian mechanism of "differentiation."[84] He argued that Darwin's approach reversed the usual method of nature by supposing that major taxonomic divisions had gradually been built up from smaller ones. No doubt evolutionary specialization did occur, but the real cause of the origin of new types was differentiation, by which Guppy meant major saltations. He now held that the production of new forms by this means had occurred as a primitive, world-ranging flora had been forced to cope with an ever-increasing diversity of climatic conditions. The evidence for this was that the oldest plant families were the most widely distributed, irrespective of varying climates or the means of dispersal. In an article with the curious title "Plant Distribution from the Standpoint of an Idealist," Guppy linked his hypothetical saltations with De Vries's mutations. The comparatively small changes observed in *Oenothera* were merely the last remnant of a much greater capacity for morphological change possessed by plants in their early history.[85] Guppy's use of the label "idealist" suggests that he was primarily concerned with using mutations as a means of preventing the identity of the basic taxa from being blurred by evolutionary speculations. Once freed from the world of the laboratory, the mutation concept could all too easily be appropriated by naturalists under the influence of more conservative anti-Darwinian values.

Slightly more influential was the work of another student of plant dispersal, J. C. Willis. Years of field experience in the tropics convinced Willis that the selection theory, with its emphasis on dispersal mechanisms and adaptation to new conditions, was not adequate to explain the geographical distribution of plants. He believed that dispersal was usually a slow process, and he argued against the evidence from the rapid spread of artificially introduced plants by pointing out that human beings had usually interfered with the native flora of areas where this occurred.[86] Adaptation also was not the key to the spread of a species, since plants seemed able to live in a great variety of climates. A more mechanistic, less opportunistic process was needed to explain the dispersal of plants, and Willis's own hypothesis was summed up by the title of his *Age and Area* (1922). He believed that the success of a family in spreading over a geographical area was quite simply a product of the length of time available to it since its original appearance. The older the family, the wider its geographical range, whatever the variation in climates to which it had been forced to adapt in the process. His evidence was the "hollow curve" obtained by plotting a graph of the number of genera against the number of species in each genus. This showed that there was a great preponderance of recent genera with only a small number of species, as opposed to a small number of older genera that had time to differentiate into many species while occupying a much wider geographical area.

Willis interpreted his evidence to suggest that each genus originated in a sudden mutation followed by a gradual dispersal during which further, smaller-scale mutations might cause it to diverge into a number of component species. Natural selection thus had nothing to do with the origin of genera; at best, it might limit their survival. Although he did not stress the world-dessication theory, he praised Guppy's work, and Guppy in turn provided a chapter in *Age and Area*. Significantly, De Vries also provided a chapter, which hailed the age and area concept as "the one great proof which the mutation theory still wanted for its acceptance."[87] Thus, De Vries himself was prepared to countenance a partial scaling up of the size of the mutations affecting evolution. By implication, Willis's views would demand saltations for the origin of all the higher taxa, although he carefully refrained from following Guppy in speculating too openly along these lines. In a later work, he again limited himself to the postulation of generic mutations, but offered some wider hints on the significance of his theory. The age and area mechanism made evolution a lawlike process, he maintained, and might be only a corollary of something even more fundamental, perhaps an electrical mechanism within life itself.[88] While recognizing that genetic mutations were caused by agents such as cosmic rays affecting the chromosomes, he criticized the small-mindedness of those who refused to consider the possibility of larger changes and hinted that there might be a definite plan of development built into the structure of living matter.[89] He also noted that Bateson had been one of the few biologists to take the age and area hypothesis seriously.[90]

Whatever the stimulus provided by the concept of genetic mutations, Guppy and Willis were forced to postulate fundamental morphological changes produced by an unknown mechanism that did not seem to be operating in their time—except, perhaps, on a very reduced scale. Guppy's self-proclaimed idealism and Willis's talk of plans built into the very nature of evolution suggest that the real inspiration of their work was an anti-Darwinian philosophy closely related to that of orthogenesis. It may be significant that Willis accepted the racial old age theory to account for the eventual decline of genera.[91] The idealists' search for law and order in evolution could be satisfied by the linear trends of orthogenesis or by the sudden production of types that preserved their identity through any change in their environment. De Vries's theory thus played a double role in the eclipse of Darwinism, as did the alternative approach to genetics favored by Bateson and Goldschmidt. Although conceived as a contribution to the new experimental study of heredity and variation, the idea of saltative evolution could all too easily be taken over by naturalists whose real sympathies lay with a more traditional way of searching for order in diversity. The geneticists were soon persuaded by their own evidence that De Vries and Bateson had overstated the case for mutations as distinct steps in evolution. Although at first their lack of field experience allowed them to ignore the influence of the environment in screening true mutations, their search for internal laws of evolution was soon purged of the quasi-mystical overtones of the idealists' ordering principle, and they were eventually able to achieve a reconciliation with the selection theory. Naturalists such as Willis, however, were opposed to Darwinism on far more fundamental grounds. Their antiutilitarianism was no mere oversight created by too close a concentration on the artificial world of the laboratory. It arose from a conviction that sudden changes under the control of some internal factor would preserve the basic orderliness of the biological groupings that Darwin's gradualism seemed to threaten. In their hands the mutation theory became, in effect, a conceptual parallel of orthogenesis, fulfilling exactly the same purpose for a different kind of phenomenon.

That some naturalists should have used the mutation theory in this way is hardly surprising, given the extent of anti-Darwinian feeling that existed in other areas of natural history. Unlike most of their colleagues, Schindewolf, Guppy, Willis, and Lotsy were dealing with phenomena that could not be interpreted in terms of conventional Lamarckism or orthogenesis, and for which a suitably modified version of Mendelism or the mutation theory provided an excellent vehicle for the expression of anti-Darwinian feelings. Despite their emphasis on discontinuity, what is ultimately most striking about their view of evolution is its emphasis on predictability, on the entire program for the evolution of life being somehow predetermined in the structure of living matter. What is surprising is not that some anti-Darwinian naturalists should have explored this alternative, but that through the work of Bateson, and perhaps the young T. H. Morgan, this conservative view of nature should have influenced the growth of what was ostensibly a new approach to biology based on experiment. What could easily have been reduced to a relatively technical debate over the size of new characters and the ability of selection to control their spread was thus exaggerated through its incorporation into the wider anti-Darwinian movement. The hostility generated in the debate with biometry was not just the product of personal rivalries that exaggerated trivial debates out of all proportion. It arose at least in part because some early supporters of both Mendelism and the mutation theory were influenced by an older and far deeper current of anti-Darwinian thinking that pushed them far beyond an overenthusiastic tendency to generalize from laboratory studies.

THE REVIVAL OF SELECTIONISM: CONCLUSIONS

Neither side regarded Willis's highly artificial bridge between genetics and natural history with any real enthusiasm, as he himself admitted in his later writings. The majority of naturalists remained loyal to gradualism and thus preferred Lamarckism or orthogenesis as a means of expressing their opposition to selection. Although made increasingly uneasy by the lack of experimental support for these alternatives, they continued to believe that the geneticists oversimplified the situation by concentrating only on processes that seemed open to laboratory study. The geneticists in turn were still prone to exaggerate the size of useful mutations and the ability of nonadaptive mutations to affect a population. Nevertheless, by 1920 some geneticists were beginning to admit a limited role for selection by conceding that useful new characters were far more likely to spread than harmful ones. What emerged from this was not exactly a new form of Darwinism, since it required the fairly rapid production of new characters through the spread of simple mutations that conferred immediate advantages upon the organism. Nevertheless, it was certainly a step in the right direction, an indication that the hostilities of the earlier debates were beginning to die down, even if the geneticists were still not yet ready to make full use of the biometricians' statistical insights.

This primitive form of selectionism can be illustrated from the work of two prominent Mendelians, R. C. Punnett and T. H. Morgan. Punnett, who wrote one of the first textbooks on Mendelism, applied the theory to the production of mimicry in butterflies in 1915. He argued from an example of polymorphism in which it could be shown that three different forms were controlled by a pair of Mendelian factors, one necessary for the manifestation of the other. Since the form with a mimetic resemblance was produced by such a simple genetic mechanism, Punnett believed that it must have appeared as a sudden evolutionary novelty. He claimed that if the protective resemblance had been built up gradually, its incipient stages would have been of no real value and could not have been aided by selection—a longstanding argument against Darwinism.[92] He did, however, admit that the character had been able to spread because it was useful, and he provided statistical tables that indicated how the reproductive advantage conferred by the character would help it to increase its frequency in the population. As a matter of fact, the tables revealed that even a very small advantage would bring about definite selective effect. Punnett ignored this and chose instead to present selection as a simple process in which new genes imparting a major advantage would soon come to dominate the population. His interpretation of mimicry was widely discussed by entomologists and thus served as a challenge to the gradualism of most field naturalists. Eventually it became the target for a specific attack by one of the founders of population genetics, R. A. Fisher, who in 1927 pointed out that Punnett had merely assumed that the phenotypic characters of the advantageous genes were always the same. Fisher showed that W. E. Castle's work on modifier genes opened up the prospect that selection acted on such modifiers to create the gradual appearance of a useful character, without affecting the basic single or double factor mechanism.[93]

In his *Critique of the Theory of Evolution* (1916), Morgan abandoned his opposition to utilitarianism, agreeing instead that a mutated gene would only be able to increase its frequency within a population if it bestowed some advantage. He was still tempted by the idea of relatively sudden changes: The appearance of eyeless forms of *Drosophila* in the laboratory suggested that blind cave animals might have lost their sight through a single mutation.[94] Morgan, however, no longer stressed discontinuity, preferring instead to present evolution as a relatively gradual process in which new genes conferring slight adaptive advantages spread into the population. He now conceded that even adaptively neutral genes would have little chance of producing any significant effect, and he explained the evolution of nonadaptive characters as a consequence of single genes having more than one effect. "One may venture the guess that some of the specific and varietal differences that are characteristic of wild types, and which at the same time appear to have no survival value, are only by-products of factors whose most significant effect is on another part of the organism where their influence is of vital importance."[95] Morgan's *Scientific Basis of Evolution* (1932) was praised by Julian Huxley because of its whole-hearted support for utilitarianism.[96] Yet Morgan's conception of selection was always somewhat oversimplified, and his colleagues have recorded the reluctance with which he faced the basic principles of Darwinism.[97] He imagined evolution proceeding not through struggle, but through the gradual incorporation of favorable new characters into the population. He could not accept the necessity for the wholesale elimination of the unfit, and actually suggested that the name "natural selection" was not really appropriate for the new explanation of adaptation.[98]

> Such a view gives us a somewhat different picture of the process of evolution from the old idea of a ferocious struggle between the individuals of a species with the survival of the fittest and the annihilation of the less fit. Evolution assumes a more peaceful aspect. New and advantageous characters survive by incorporating themselves into the race, improving it and opening to it new opportunities. In other words, the emphasis may be placed less on the competition between the individuals of a species (because the destruction of the less fit does not *in itself* lead to anything that is new) than on the appearance of new characters and modifications of old characters that become incorporated into the species, for on these depends the evolution of the race.[99]

In Morgan's view, selection meant only that unfavorable mutations were prevented from breeding, not that significant numbers of the unfit must be eliminated in order to drive the process in the right direction. Evolution occurred only when mutation presented something new to the environment, and each mutation was dealt with as and when it occurred. This was still evolution without the unpleasant implications of the old Darwinism.

Morgan's view of selection may have been oversimplified, but he had conceded the basic point that utility did determine the success or failure of new characters. Many field naturalists still regarded selection in any form as unacceptable—the position adopted in Robson and

Richards's massive 1936 survey of animal variation.[100] The prevalence of nonadaptive characters was still widely accepted as a major argument against the selection theory.[101] Even among those more sympathetic to selection, there was still a temptation to look for loopholes that would allow room for the evolution of nonadaptive characters. Julian Huxley's mechanism of allometry, based on differential growth rates, has already been cited as a means whereby the nonadaptive trends of orthogenesis could be accommodated. In 1927 the ecologist Charles Elton suggested that nonadaptive genes could take hold in a population during periods when the intensity of selection was reduced; for instance, when numbers increased after illness or a drought.[102] Most naturalists were still convinced of the effectiveness of anti-Darwinian mechanisms such as Lamarckism and orthogenesis. At a conference held at Tübingen in 1929, there was complete disagreement between the naturalists and the geneticists, yet over the next decade the situation was to change dramatically. A new and revised form of selectionism rapidly captured the loyalty of many naturalists and allowed them to accept genetics as a basic component of evolution theory. At another conference held at Princeton in 1947 there was an almost embarrassing lack of discussion, so great was the success of the new Darwinian theory.[103]

What had happened to alter the climate of opinion so drastically? The simple answer is that what we now know as the "modern synthesis" had emerged, although historians are only just beginning to clarify the steps in this process. Most obviously, the creation of a new science of mathematical population genetics allowed the formulation of theoretical models in which the action of selection upon small genetic effects played a crucial role. This was the work of R. A. Fisher and J.B.S. Haldane in England and Sewell Wright in America. At last the mathematical sophistication of biometry was applied to a genetical analysis of the population to show that the old Darwinian processes could indeed be interpreted in Mendelian terms. Selection acted not on single genes as they were created by mutation, but on a gene pool, which constituted the species' fund of variability, constantly replenished by mutation and genetic recombination. Equally important, though, was the work of those naturalists who showed that the new selectionism could be applied to the various areas of study, thus eliminating the old anti-Darwinian arguments and demonstrating that here was a mechanism of evolution capable of giving fruitful insights to guide field research, yet compatible with the laboratory biologists' understanding of variation and heredity. Naturalists such as Julian Huxley, Theodosius Dobzhansky, and Ernst Mayr, and paleontologists such as George Gaylord Simpson, paved the way for scientific disciplines to accept the new approach and thereby precipitated a major revolution in evolution theory.

The reemergence of Darwinism was thus the result of two separate processes of reconciliation. First, the gulf between biometry and Mendelism had to be bridged. From the beginning it had been clear that discontinuous variation was not incompatible with natural selection. In his own way De Vries had demonstrated this point, although far more significant in the long run was G. Udney Yule's early suggestion that a series of Mendelian factors affecting a single character would provide the apparently continuous range of variation studied by the Darwinians. The heat of the original debate was so intense that Yule's point was ignored, so that it required the emergence of a new generation of biologists trained in statistics to appreciate its potential significance. Fisher provides perhaps the best example of someone trained in Pearson's techniques, yet ready to brave his mentor's anger in order to explore this new possibility for selectionism.

After the creation of a science of population genetics, it was then necessary to reconcile the hereditarianism of the Mendelian and Darwinian positions with the preference of many field naturalists for mechanisms of Lamarckism and orthogenesis, both of which denied the isolation of the germ plasm. This task was rendered easier by the fact that the naturalists themselves were only too well aware of the lack of experimental verification for these beliefs. As long as the genetical selection theory had seemed crude and oversimplified, the absence of an experimental foundation had not been so crucial; but once it could be demonstrated that a new form of selectionism had emerged, consistent with the laboratory work and yet capable of guiding field research in a productive way, support for the alternatives waned rapidly. The failure of Lamarckism and orthogenesis to adapt themselves to the new biology of the twentieth century had given rise to frustration and even cynicism toward evolution theory among these remaining, rather lukewarm supporters of the anti-Darwinian mechanisms. This was just the kind of atmosphere that would respond to a major initiative offering a way out of the impasse. The genetical selection theory succeeded because the only possible source of opposition to it was based upon conceptual foundations of early nineteenth-century origin that had never—despite numerous attempts—been translated into the terms of twentieth-century science.

Darwinism triumphed in the end because a reconciliation with Mendelism was not only possible but inevitable if laboratory and field workers were not to remain permanently estranged. At the same time, the rise of genetics necessarily undermined the alternative view of organic development underlying the other anti-Darwinian theories. This pinpoints what is probably the most important conclusion emerging from my study: that the eclipse of Darwinism was a hybrid event stemming from the coincidence in time of two developments in the history of biology that had only one thing in common—an anti-Darwinism that turned out to be entirely superficial on one side. The great wave of dissatisfaction with the selection theory at the end of the nineteenth century generated two movements, one essentially forward-looking, the other backward. The theories of Lamarckism and orthogenesis represented the revival of an earlier, largely teleological view of development in which the evolution of life was modeled on the goal-directed process of individual growth. Concepts that were originally identified with a pre-Darwinian, nonevolutionary world view were modified to eliminate their more obviously teleological implications so that they could serve as a scientific alternative to the more rigorous materialism of the selection theory. Whatever the progressive social implications drawn from Lamarckism, the theory's scientific basis was an attempt to create a viable evolutionary mechanism that would preserve certain conservative intellectual values drawn from natural theology and idealism. Parallel with this movement, however, was a more rigorous attempt to reform biology, which used the excuse of Darwinism's apparent exhaustion to extol the necessity of studying the underlying processes of heredity and variation with the techniques of the laboratory. Although Lamarckism and orthogenesis tried to make older concepts scientific as measured by the standards of natural history, genetics and the mutation theory arose out of a new definition of what a scientific approach to evolution theory ought to be. These new theories did not share the other alternatives' commitment to an orderly vision of evolution modeled on ontogeny. Instead, they accepted a rigid hereditarianism that linked them, willingly or not, with Weismann's neo-Darwinism. Their opposition to selection may have been stimulated by anti-Darwinian attitudes imported from the more traditional alternatives, but in general it rested on an overenthusiastic desire to stress new insights at the expense of old ones rather than on any fundamental conceptual incompatibility.

If this interpretation is valid, it could be argued that the real eclipse of Darwinism was caused by the resurgence of the more conservative philosophy of development, while the debates with Mendelism and the mutation theory were merely the growing pains of the metamorphosis that was needed to transform Darwinism into an effective mechanism of evolution. Whether stressing the purposefulness of adaptation through Lamarckism or the orderliness of nonadaptive development through orthogenesis, the earlier philosophy rejected the fundamental insights of Darwinism by transferring the driving force of evolution to an internal level where it partook of the teleological nature of individual growth. Had the experimental movement not been contaminated by extreme anti-Darwinian sentiments derived from this more traditionally oriented philosophy, the emergence of modern genetics might have been accompanied by a far less obvious repudiation of selection. As it turned out, the split that did occur was mended within a few decades, while the rise of genetics spelled the end of the recapitulation theory and the mechanisms of evolution based upon it. The hereditarian viewpoint of genetics was incompatible with the belief that the purposeful characters of individual growth could somehow be fed into the evolutionary process linking successive generations. The supporters of Lamarckism and orthogenesis did their best to adapt their theories to the new experimentalist methodology, but were never very successful. Their fascination with ontogeny led them to stress the experimental study of embryology, thereby leaving them with no real defense against the takeover of inheritance theory by genetics. Experimental Lamarckism failed, and the theory at last retained a hold only within those disciplines that were least affected by the new biology.

In the end, perhaps the most surprising thing about the eclipse of Darwinism is that it took the idealist view of development so long to present itself as an alternative basis for evolutionism. This must have been due to a number of factors associated with the way in which scientists perceived the implications of the theories available to them. It has been said that evolutionism was "in the air" in the mid-nineteenth century, but this is only partly true. The old view of nature was capable of generating a developmental outlook, but it was so heavily steeped in an essentialist concept of organic form that it could not serve as the basis for a fully articulated mechanism of natural evolution. That is why Darwinism played an essential part in breaking the deadlock by confirming once and for all that a naturalistic theory of evolution was possible. Darwinism then capitalized on its pioneering role by adopting a flexible position that allowed some room for alternative mechanisms and thus shut out much of the original criticism. It was inevitable that the naturalists who remained committed to the older philosophy would try to regain their position by modernizing their viewpoint as an alternative to Darwinism, but it took some time for them to reorganize themselves. Their immediate temptation was to try out a theistic form of evolutionism that was too obviously a compromise with the old idea of a divine plan of creation. Only gradually did they face up to the necessity of fully purging the element of divine control from their thoughts to create an acceptably naturalistic theory. Neo-Lamarckism thus arose—not out of a revival of Lamarck's own ideas, which had been thoroughly discredited by an earlier generation—but out of the theistic evolutionists' desire to provide a more natural explanation of adaptation. Orthogenesis was again a product of their efforts to impose the orderliness of ontogeny upon evolution without seeming to imply a progressive goal toward which life was programed to advance.

These alternatives gained a chance to dominate the scientific community because they were formulated just as the original Darwinism was running out of steam. Given the immense limitations of its ideas on heredity and the wild expectations of a complete reconstruction of the history of life from an incomplete fossil record, it was inevitable that the theory would soon reach the limit of its ability to guide fruitful research. This gave its opponents the chance to exploit a wide range of existing anti-Darwinian arguments in order to demand a wholesale reexamination of the basic points of evolution theory. The Darwinians did not help matters by setting up a more dogmatic version of their theory that seemed even more vulnerable to its opponents' claims. Yet both biometry and Weismann's concept of the germ plasm held potential for future development, though not in the form in which they were first presented. The experimental study of variation and heredity was a different kind of response to Darwinism's exhaustion, one that promised a positive solution to some of its problems that would preserve its key insights. Since it retained the hereditarianism of Weismann's position, genetics did not desert the central element of the Darwinian philosophy, although the dispute over the level of discontinuity in variation was blown up into a major debate. Genetics and the mutation theory at first seemed to reinforce the decline of Darwinism, thus adding to the confusion existing in early twentieth-century evolutionary thought. In the process, however, they steadily undermined the Lamarckian alternative and made it possible for scientists to create a new synthesis with the selection theory.

It would be easy to say that Lamarckism collapsed because its basic idea was false, and that scientists were bound eventually to realize this; but we cannot explain the failure of the anti-Darwinian movement in terms of truth and falsehood alone. Ted Steele's recent attempts to reformulate Lamarckism along lines acceptable to modern genetics, and to provide it with experimental evidence, show that the basic concept is not quite so obviously untrue. In any case, we cannot use the falsity of Lamarckism to explain its fall without invoking wider factors to explain its original rise. In fact, both the rise and fall of the Lamarckian alternative must be seen in terms of wider developments occurring in evolution theory, in biology as a science, and in society as a whole. Lamarckism did help to sustain certain religious and philosophical values, and it was useful in the ideological battle against extreme laissez-faire economic policies; but this does not mean that it arose solely out of wishful thinking. The crisis in Darwinism was real enough, while Lamarckism's own confrontation with experimentalism would only become acute after 1900. There can be little doubt that genetics itself drew some of its vitality from the backlash of hereditarian attitudes that created the eugenics movement. Lamarckism collapsed not only as a science, but also as a cultural movement. Its experimental failure was reinforced by a massive desertion of all those forces from outside science that had once sustained it. The thread of literary antimaterialism that linked Bernard Shaw to Arthur Koestler preserved only selected items from a Lamarckian tradition that had become defunct as an intellectual force even before its scientific failings were exposed.

In its original form, then, neo-Lamarckism is dead. Even Steele's modern interpretation of the theory takes for granted certain basic components of the genetic scheme and certainly does not invoke the quasi-mystical guiding role given to ontogeny by its earlier supporters. The basic idea of the inheritance of acquired characters may very well be able to exist independent of the theoretical framework that once supported it, and perhaps it may still be able to uphold some of the broader implications that writers such as Koestler have continued to draw from it. The rise and fall of Lamarckism represents not the testing and refutation of a simple biological possibility, but the revival of a whole complex of traditional beliefs about nature and the eventual demonstration of their incompatibility with modern attitudes and information. Furthermore, it must be noted that for all its progressive and humanistic aspects, this complex rested upon a profoundly conservative intellectual foundation. Considering that it also supported the theory of racial senility and a hierarchical view of organic relationships that was a major force in the establishment of race theory, modern Lamarckians should be grateful that it was eliminated. If modern biology ever does concede a place for the inheritance of acquired characters, the concept will have to be interpreted in a manner very different from that in which it first flourished. Opponents of the modern selection theory who dismiss it as an expression of blind materialism or capitalist ideology should pause for a moment to think that it it were not for the triumph of the modern synthesis, their own alternatives might still be burdened with equally distasteful implications.

Most biologists still feel that the chances of a real Lamarckian revival are pretty slim. A far more intensive debate has raged in recent years over the question of continuity in evolution, a conflict sparked by the suggestion of Stephen Gould and Niles Eldredge that the process works through "punctuated equilibria."[104] Supporters of this theory try to minimize the extent of their disagreement with Darwinism, but opponents dismiss it as little more than a return to the old form of the mutation theory as advocated by De Vries. Certainly, the "punctuations" of this theory need not involve macromutations—although its supporters have perhaps been more willing than most biologists to admit that saltations might play a role in evolution. To a large extent, the new approach derives from the orthodox Darwinian belief that speciation will take place most readily when small populations are confined in isolated areas under extreme conditions. Under these circumstances, change of form will be fairly rapid, and the chances of it being recorded by a sequence of fossils will be slight. Only when a new species spreads out over a wide area, following the extinction of the parent form, will it appear quite suddenly in the fossil record. Speciation is a rapid process by geological standards, although the punctuationists insist that this does not mean that it is based on sudden genetic changes.

Because punctuated equilibrium theory accepts that speciation occurs in peripheral areas, it holds that the range of new species will represent a kind of random variation with respect to any evolutionary trends affecting the group. Such a trend will be a consequence of the differential success or failure of the new species, a process of species selection in which the species play the same role as individuals in the normal form of natural selection. This does indeed seem reminiscent of De Vries's own interpretation of the mutation theory, although the punctuationists have a less drastic explanation of how the new species are actually produced. The crucial point is that the paleontologists who support the punctuated equilibrium model insist that the fossil evidence cannot, after all, be reconciled with gradual evolution. Once a species has established itself, it remains essentially static until it is replaced. The orthodox Darwinian mechanism of evolution by the selection of individual variations is thus discredited, leaving the comparatively rapid speciation-events as the only source of new characters.

This rejection of the modern Darwinian synthesis has very broad implications. The puntuationists doubt that the rigid stability of well-established species can be maintained solely by the action of natural selection in eliminating individuals who vary too far from the norm. They suspect that a more powerful force must be at work, perhaps a constraint imposed by the existing course of individual development.

Such constraints would arise from the fact that ontogeny has to follow a viable pattern, with which many random mutations would not be compatible. Speciation may take place when the constraints are dismantled under exceptional circumstances, allowing rapid change toward a new pattern of development. If this is the case, then the production of new characters during speciation may not be entirely random. The existing pattern of development may only allow change in certain directions, thereby producing an origin-bias that directs evolution independent of selection. This bias has been compared to the concept of mutation-pressure invoked by some early geneticists, but the appeal to ontogeny as a guiding force in evolution seems more reminiscent of orthogenesis. Under the name "epigenetic evolution," some biologists are once again exploring the possibility that individual growth has a logic of its own that may exert some control over the production of new characters in evolution.[105] The supporters of this new approach have no desire to revive the old idea of a mystical guiding force, nor do they suppose that massively nonadaptive characters could be produced in this way. Nevertheless, the renewed interest in ontogeny shows that certain aspects of the earlier anti-Darwinian viewpoint may have had some validity.

For those biologists who take it seriously, epigenetics has made the idea of genetically sudden changes plausible once again. The creation of a viable new pattern of development may result from a small mutation affecting the very early stages of growth, thereby switching the process into a new direction. A few now believe that such changes represent the only source of new characters, and they explicitly identify themselves with the long-neglected views of Goldschmidt and Schindewolf. The "hopeful monster" has now returned to evolution theory. Even if these sudden changes are not the basis of all speciation events, they may be the source of those major new initiatives in the direction of evolution that lead to the formation of new types or classes. In this case, the Darwinian emphasis on adaptation may be misguided, since the character of a viable developmental path may be determined by purely internal factors. Once a fundamentally new structure has appeared, it will be exploited by adaptive evolution, but the basic pattern of the structure may be determined by deeper forces that do not depend upon selection. Here again we see elements of an older anti-Darwinian tradition resurfacing in modern biology. Despite accusations that the renewed support for discontinuous evolution stems from the influence of Marxism, there can be little doubt that the new program also has roots running back to an older, and originally far more conservative, way of thought.

This is not meant to imply that the new ideas must be tainted with the outdated teleology that led the last generation of biologists to dismiss orthogenesis as unscientific. The first attempt to displace the selection theory failed because it was still based on the assumption that nature must unfold in an orderly pattern that was somehow predetermined. The goal-directed process of individual growth was almost inevitably seen as the most likely means by which this kind of order could be imposed. The teleological aspects of this assumption were concealed to some extent, but not eliminated; and the theory of orthogenesis was rejected as incompatible with Mendelian genetics. The modern alternatives to Darwinism, however, accept the central dogma of molecular biology and do not suppose that the growth of the individual can point out trends that evolution must follow in the future. By treating epigenetics as a factor that can limit or channel the potentialities introduced by genetic mutations, the traditional concern for the orderliness of growth can be expressed in a way that does not violate the accepted principles of causality. It is also claimed that by appealing only to small mutations affecting the early stages of growth, the new approach avoids the pitfalls that were revealed by modern population genetics. The hopeful monster, which appears as a single individual with no breeding companions, is an evolutionary dead end; and it was for this reason—as well as the general suspicion of large mutations—that Goldschmidt's views were ignored. By invoking small mutations within an isolated population, this difficulty may be overcome. It is certainly conceivable that in the excitement generated by the synthesis of Mendelism and natural selection, some genuine problems may have been overlooked. The few scientists who argued that apparently discredited ideas contained a kernel of truth that would help to solve these problems were ignored until recently. Some modern biologists now believe that an even broader synthesis of the various modes of change is possible. The potential of their new approach remains to be determined, of course, and there are still many skeptics who believe that the challenge to modern Darwinism is unnecessary.

The greatest threat to modern evolutionism comes from those who proclaim that the enterprise of trying to reconstruct the development of life on earth is fundamentally misguided. Until recently, it would have been unthinkable that such an opinion could flourish within science itself, yet a few working biologists have now taken up this position. They are the most radical exponents of "cladism," a new and more precise technique for classifying living forms, which they insist does not permit the identification of evolutionary relationships. Hence, they argue, all efforts to reconstruct the evolutionary succession of forms have been based on mere guesswork, with no hope of scientific confirmation. It is also argued that the hypothetical adaptive forces invoked to explain particular episodes of evolution are completely untestable. Such views do seem to parallel those of the most radical early Mendelians, although then, as now, the majority of biologists continued to believe that reconstruction of the way evolution worked was a valid scientific problem. The Mendelians were in any case soon distracted by the question of the evolutionary mechanism itself. The modern cladists are not in a position where they are likely to be diverted in this way, although their critics insist that with further refinement their techniques may yet turn out to be of relevance to evolutionism.[106] In general, it would seem that whenever conflict arises over the mechanism of evolution, a few more extreme critics will begin to argue that the whole concept of evolution is unworkable as science. The future success of the cladists' position may well depend upon whether or not the current debates over the mechanism of evolution are resolved to the satisfaction of most biologists.

Finally, what of that most vociferous source of opposition to evolutionism from outside the ranks of orthodox biology; creationism? Although creationists are always ready to seize upon anything that offers the chance of making a point against Darwinism, a study of the scientific opposition to the selection theory generates more problems than opportunities for their program. In one sense, they are already familiar with many of the points made during the earlier debates, since their own arguments against selection often reflect the simpleminded image of that theory prevalent during the early twentieth century. Nevertheless, the eclipse of Darwinism shows that the creationists' attempt to portray natural selection and divine creation as the only alternatives worthy of equal time in the schools is a gross distortion of the real situation in science. We have seen that a massive wave of opposition to Darwinism within biology did not in general create doubts about the basic idea of evolution. On the contrary, a wide range of alternatives was proposed, some of which may yet be revived as part of a new evolutionary synthesis. Are all of these alternatives to be awarded equal time in the schools, along with the more esoteric challenges to scientific orthodoxy by Velikovsky and the like? It is not by accident that my survey of the eclipse of Darwinism has been completed without reference to the outburst of fundamentalist opposition to the teaching of evolution that culminated in the famous "monkey trial" of John Thomas Scopes in 1925. By that time, support for Darwinism was already reviving in some areas of biology, although Henry Fairfield Osborn—who was anything but an exponent of the selection theory—

played a leading role in organizing the scientific support for Scopes. The scientific challenge to Darwinism was certainly influenced by religious and moral factors, but distrust of Darwinian materialism led the critics toward Lamarckism, not creationism.

Modern creationists have seized upon the recent scientific debates as evidence that Darwinism is about to collapse, leaving their own position as the only reasonable alternative. Far from heralding the imminent collapse of a discipline, however, willingness to propose and discuss new hypotheses is generally taken as a sign of scientific vitality. Nevertheless, the creationists have gained some comfort from those more extreme scientific critics who have declared that the whole evolutionary approach is unworkable. If we regard all studies of the past development of living forms as outside the scope of science, then evolution and creation are both reduced to the same level. The refusal of many scientists to accept a position in which all questions about the past become unanswerable is a function of their belief that such a position gives too much away to those who attack evolution not on the grounds of scientific caution, but because they wish the whole question transferred into the realm of the supernatural, where their own religious views can be given full control.

In the end, the real challenge of modern creationism goes far deeper than the attacks on evolutionism that hit the headlines. Since the position of most creationists is that the world was created almost instantaneously only a few thousand years ago, their quarrel is not just with evolution theory, but with the entire scientific interpretation of the past. Geology, paleontology, archaeology, and even cosmology must all be rewritten if this position is to be maintained, thus returning science to a theoretical framework that has not been taken seriously since the eighteenth century. In these circumstances, evolution would have to be abandoned almost as a casual by-product of the reduction in the time span of earth history. Whatever their views on Darwinism, few scientists see even the remotest possibility of such a development taking place. If the creationists' scientific counterrevolution succeeds, it will produce an intellectual upheaval of such magnitude that any eclipse of Darwinism would seem trivial by comparison.

NOTES

CHAPTER 1

The complete reference for each item cited will be found in the Bibliography.

1. Salisbury, Presidential Address.

2. For a detailed discussion of this issue see Joe D. Burchfield, *Lord Kelvin and the Age of the Earth.*

3. Henry Fairfield Osborn's account of the debate is quoted in Leonard Huxley, *Life and Letters of Thomas Henry Huxley*, 3: 326–27.

4. Alfred Russel Wallace, "The Problem of Utility."

5. E. Ray Lankester, "Are Specific Characters Useful?" See also the letters under the title "The Utility of Specific Characters" by J. T. Cunningham, W. T. Thistleton-Dyer, and W. F. R. Weldon.

6. George John Romanes, *Darwin and After Darwin;* Ludwig Plate, *Über Bedeutung und Tragweite des Darwin'schen Selectionsprincips* and the more detailed works of 1903 and 1913; and Vernon L. Kellogg, *Darwinism Today.* See also Yves Delage and Marie Goldsmith, *The Theories of Evolution.*

7. Eberhart Dennert, *At the Deathbed of Darwinism*, a translation of *Von Sterbelager des Darwinismus.*

8. Julian Huxley, *Evolution: The Modern Synthesis*, pp. 22–28.

9. Koestler's *Case of the Midwife Toad* reconstructs the events leading up to the discrediting of Paul Kammerer's Lamarckian experiment on the midwife toad. Koestler's general support for Lamarckism can be found in *The Ghost in the Machine* and *Janus: A Summing Up.*

10. Loren Eiseley, *Darwin's Century*, chap. 9 discusses Hugo De Vries's mutation theory.

11. Phillip Fothergill, *Historical Aspects of Organic Evolution*, chaps. 6–10.

12. Erik Nordenskiöld, *The History of Biology*, part 3. See, for example, the final page (616) for Nordenskiöld's own doubts about Darwinism.

13. See, for instance, William B. Provine, *The Origins of Theoretical Population Genetics.* For more detailed references see below, Chaps. 2 and 7.

14. Considerable excitement has been generated in the scientific literature by the work of E. J. Steele; see his *Somatic Selection and Adaptive Evolution.*

15. Stephen Gould, "The Eternal Metaphors of Paleontology."

16. Kuhn draws most of his examples from revolutions in the physical sciences, but does refer briefly to the Darwinian revolution. See *The Structure of Scientific Revolutions*, pp. 171–72, where he makes the important point that the real issue involved was the destruction of the old belief in a harmoniously structured universe.

17. Ernst Mayr,"The Nature of the Darwinian Revolution," in Mayr, *Evolution and the Diversity of Life,* pp. 277–96, and John C. Greene, "The Kuhnian Paradigm and the Darwinian Revolution in Natural History." Both writers, however, deal more with complexities in the situation leading up to the appearance of Darwin's theory.

18. David Hull, "Sociobiology: A Scientific Bandwagon or a Traveling Medicine Show?"

19. James R. Moore, *The Post-Darwinian Controversies.*

20. Richard Hofstadter, *Social Darwinism in American Thought.*

21. Robert C. Bannister, *Social Darwinism.* On eugenics see, for instance, Hamilton Cravens, *The Triumph of Evolution.*

22. J. B. S. Haldane, *Heredity and Politics.* The other British pioneer of the genetical selection theory, R. A. Fisher, was a firm supporter of eugenics—although he was well aware of the complexities of the situation.

23. Hofstadter recognizes the Lamarckian influence on Lester Ward; see *Social Darwinism in American Thought,* chap. 4. See also George Stocking, "Lamarckianism in American Social Thought," and Lester G. Stevens, "Joseph LeConte's Evolutionary Idealism."

24. John S. Haller, *Outcasts from Evolution,* chap. 6.

25. Stephen Gould, *Ontogeny and Phylogeny,* chaps. 4 and 5.

CHAPTER 2

1. On Chambers see Milton Millhauser, *Just Before Darwin,* and M. J. S. Hodge, "The Universal Gestation of Nature." See also Bowler, *Fossils and Progress,* chap. 3.

2. The literature on the origins of Darwin's theory is too extensive to be cited in detail here. Standard works include Sir Gavin de Beer, *Charles Darwin;* Loren Eiseley, *Darwin's Century;* Michael T. Ghiselin, *The Triumph of the Darwinian Method;* and John C. Greene, *The Death of Adam.* More recently, much has been written in the light of the unpublished Darwin papers at Cambridge University Library; see, for instance, Howard E. Gruber and Paul Barrett, *Darwin on Man;* Edward Manier, *The Young Darwin and his Cultural Circle;* and Sylvan S. Schweber, "The Origin of the *Origin* Revisited."

3. I say "somewhat similar," having argued elsewhere that Wallace did not have the complete theory of natural selection in 1858; see "Alfred Russel Wallace's Concepts of Variation." Since writing this article I have noticed the same point made in the conclusion of Henry Fairfield Osborn's *From the Greeks to Darwin,* although Osborn does not analyze Wallace's paper for proof. This is not meant to deny Wallace's important contributions to the later development of the selection theory.

4. On the distinction between the utilitarian and idealist concepts of design, see Bowler, "Darwinism and the Argument from Design." See also Thomas McPherson, *The Argument from Design.*

5. For a survey of the discussions on Darwinism and progression, see Bowler, *Fossils and Progress,* chap. 6.

6. For general surveys of the Darwinian debate see, for instance, John C. Greene, *The Death of Adam;* Loren Eiseley, *Darwin's Century;* Alvar Ellegård, *Darwin and*

the General Reader; Peter Vorzimmer, *Charles Darwin: The Years of Controversy;* and David L. Hull, *Darwin and His Critics.*

7. Hugh Miller, *Footprints of the Creator.* See *Fossils and Progress,* chap. 4.

8. Kelvin was prepared to allow at most one hundred million years, far less than Darwin had supposed. See his "On Geological Time" (1868), reprinted in Kelvin, *Popular Lectures,* 2: 10–72.

9. Nägeli's argument was proposed in his *Entstehung des Begriffs der naturhistorischen Art* and developed further in the *Mechanische-physiologische Theorie des Abstammungslehre* of 1884. A related argument advanced by Mivart was that useful organs must pass through incipient stages even when they can have been of no real value.

10. Darwin, *Origin of Species,* 6th ed., pp. 170–76.

11. St. George Jackson Mivart, *On the Genesis of Species,* pp. 84–87. On Mivart's relations with Darwin, see Jacob W. Gruber, *A Conscience in Conflict,* and Vorzimmer, *Charles Darwin: The Years of Controversy,* pp. 225–51.

12. Fleeming Jenkin, "The Origin of Species." Loren Eiseley suggests that Jenkin's argument forced Darwin to retreat from selectionism altogether; see *Darwin's Century,* pp. 209–16. However, it has been shown that the argument was crucial for sports, not for normal variation; see Peter Vorzimmer, "Charles Darwin and Blending Inheritance," and *Charles Darwin: The Years of Controversy.* See also Bowler, "Darwin's Concepts of Variation."

13. Erik Nordenskiöld, *The History of Biology,* p. 528.

14. Alvar Ellegård, *Darwin and the General Reader,* pp. 272–73.

15. See especially Huxley's "On the Reception of the 'Origin of Species,'" in Francis Darwin, ed., *The Life and Letters of Charles Darwin,* 2: 179–204.

16. Darwin, *Descent of Man,* p. 61. See, for instance, Eiseley's *Darwin's Century,* p. 209, which refers to this passage and attributes the retreat to Jenkin's attack—although Darwin mentions Nägeli's criticisms in particular. Cf. note 9 above.

17. Darwin, *Origin of Species* (1859), pp. 132–39 and 159.

18. See, for instance, T. H. Huxley's "Evolution in Biology," reprinted in his *Collected Essays,* vol. 2, *Darwiniana,* 187–226, p. 223.

19. Michael Bartholomew, "Huxley's Defence of Darwin."

20. See Wallace, "The Limits of Natural Selection as Applied to Man," reprinted from *Contributions to the Theory of Natural Selection* (1870) in *Natural Selection and Tropical Nature,* pp. 186–214.

21. See the papers collected in Gray's *Darwiniana.*

22. Note the subtitle of Haeckel's *History of Creation. (A Popular Exposition of the Doctrine of Evolution in General and that of Darwin, Goethe and Lamarck in Particular.)*

23. Weismann, *Studies in the Theory of Descent.*

24. Michael T. Ghiselin, *The Triumph of the Darwinian Method.* Darwin's researches were published in works such as *The Various Contrivances by which Orchids are Fertilized by Insects, Insectivorous Plants,* and *The Power of Movement in Plants.*

25. Henry Walter Bates, "Contributions to an Insect Fauna of the Amazon Valley." Parts of this paper are reprinted in Barbara G. Beddall, *Wallace and Bates in the*

Tropics, pp. 213-17. On Bates's travels, see his *The Naturalist on the River Amazons* and George Woodcock, *Walter Bates: Naturalist of the Amazons.*

26. Julian Huxley, *Evolution: the Modern Synthesis,* pp. 412-17 and 457-66.

27. See Edward B. Poulton, *The Colours of Animals* and his collected papers, *Essays on Evolution.* In the introduction to the latter work, Poulton criticizes Bateson for his lack of field experience; see pp. xiii-xlviii.

28. See Wallace, "On the Zoological Geography of the Malay Archipelago"; *The Malay Archipelago;* and *Island Life.* On Wallace's line, see Beddall, *Wallace and Bates in the Tropics,* pp. 143-59; Martin Fichman, "Wallace: Zoogeography and the Problem of Land Bridges"; and Ernst Mayr, "Wallace's Line in the Light of Recent Zoogeographical Studies," reprinted in *Evolution and the Diversity of Life,* pp. 626-45.

29. I have used the version of Hooker's essay reprinted in the *American Journal of Science:* "On the Origination and Distribution of Vegetable Species."

30. See the papers "Species as to Variation, Geographical Distribution and Succession" and "Sequoia and its History: The Relations of North American to Northeast Asian and Tertiary Vegetation," reprinted in Gray, *Darwiniana,* pp. 146-68 and 169-95.

31. Wallace, "On the Phenomena of Variation and Geographical Distribution as Illustrated in the Papilionidae of the Malayan Region." This was reprinted under the title "The Malayan Papilionidae or Swallow-Tailed Butterflies . . ." in Wallace's *Contributions to the Theory of Natural Selection,* pp. 130-200, but was omitted from *Natural Selection and Tropical Nature* because he thought it was too technical.

32. See Ernst Mayr, "Darwin and Isolation" and "Darwin, Wallace, and the Origin of Isolating Mechanisms," both reprinted in *Evolution and the Diversity of Life,* pp. 120-28 and 129-34.

33. See Romanes, "Physiological Selection: An Additional Suggestion on the Origin of Species" and *Darwin and After Darwin,* 3, e.g., pp. 144-45. On Romanes's theory see John E. Lesch, "The Role of Isolation in Evolution."

34. H. M. Vernon, "Reproductive Divergence: An Additional Factor in Evolution."

35. Moritz Wagner, *Die Darwin'sche Theorie und das Migrationsgesetz der Organismen,* and the English translation. Wagner's collected papers appeared in 1889 as *Die Entstehung der Arten durch räumliche Sonderung.*

36. On Darwin's reaction, see the essays by Mayr cited above in n. 32. Weismann's attack is his *Über den Einfluss der Isolierung auf die Artbildung.*

37. John Thomas Gulick, "Divergent Evolution through Cumulative Segregation." On Gulick's work see Addison Gulick, *Evolutionist and Missionary: John Thomas Gulick,* and John E. Lesch's "The Role of Isolation in Evolution." Romanes himself regarded Gulick's work as important without abandoning physiological selection; see *Darwin and after Darwin,* 3: 11-24. Note also Kellogg's favorable reaction, *Darwinism Today,* pp. 249-53.

38. Mayr, "Karl Jordan on Speciation," "Karl Jordan on the Theory of Systematics and Evolution," and "Karl Jordan and the Biological Species Concept," reprinted in *Evolution and the Diversity of Life,* pp. 135-43, 297-306, and 485-92. Jordan's most important paper is his "Der Gegensatz zwischen geographischer und nichtgeographischer Variationen," but see also his attack on Vernon's ideas: "Reproductive Divergence not a Factor in the Evolution of New Species."

39. David Starr Jordan, "The Origin of Species through Isolation." Another favorable article is F. W. Hutton, "The Place of Isolation in Organic Evolution."

40. See Rudwick, *The Meaning of Fossils*, pp. 245–49, which refers to Albert Gaudry's *Animales fossiles et géologie de l'Attique . . .* (1862–67).

41. See Rudwick, p. 250, and Bowler, *Fossils and Progress*, pp. 132–33. Marsh's articles on the Tertiary animals are listed in the Bibliography. On his life and work, see Charles Schuchert and Clara Mae Levene, *O. C. Marsh: Pioneer in Paleontology*.

42. See, for instance, Marsh, *Dinocerata*, p. 59, and for a discussion, Bowler, *Fossils and Progress*, p. 137.

43. Rudwick, *The Meaning of Fossils*, pp. 255, 260. Note in contrast Russell's claim that the preference of evolutionists for explaining developments in terms of changing old organs rather than forming totally new ones stemmed from a Darwinian reluctance to attribute a creative force to nature; see *Form and Function*, pp. 306–7.

44. Huxley, "Remarks on *Archaeopteryx lithographica*" and "On the Animals which are most nearly Intermediate between Birds and Reptiles," reprinted in *Scientific Memoirs*, 3: 340–45 and 303–13. See Rudwick, *The Meaning of Fossils*, pp. 249–50 and Bowler, *Fossils and Progress*, pp. 134–35.

45. See Marsh's articles listed in the Bibliography and his monograph, *Odontornithes*.

46. Russell, *Form and Function*, pp. 357–58. See Depéret, *Les transformations du monde animal*, and Zittel, "Palaeontology and the Biogenetic Law."

47. Russell, *Form and Function*, pp. 246–67.

48. See Gould, *Ontogeny and Phylogeny*, chap. 1.

49. Fritz Müller, *Für Darwin*, translated as *Facts and Arguments for Darwin*; see pp. 111–12.

50. Gould notes, for instance, Arthur O. Lovejoy's confusion of this issue in the very title of his "Recent Criticism of the Darwinian Theory of Recapitulation . . ."

51. Gould, *Ontogeny and Phylogeny*, pp. 102–9.

52. See Russell, *Form and Function*, pp. 260–67.

53. Marsh, *Dinocerata*, p. 173. See also T. H. Huxley's "Paleontology and Evolution" (1868), reprinted in his *Essays*, 8: 340–88, pp. 362–64.

54. Haeckel, *History of Creation*, 2: 247–48. Huxley had suggested this interpretation in his (unpublished) Hunterian lecture for 1866. On the modern view that the animals are indeed polytypic, see George Gaylord Simpson, "Mesozoic Mammals Revisited."

55. Huxley, "On the Characters of the Pelvis in the Mammalia, and the Conclusions Respecting the Origins of the Mammals which may be Based on them" (1869), reprinted in *Scientific Memoirs*, 4: 345–56.

56. Seeley published a long series of articles under the collected title "Researches on the Structure, Organization and Classification of the Fossil Reptiles . . ." Important individual articles are cited in the Bibliography; see William E. Swinton, "Harry Govier Seeley and the Karoo Reptiles."

57. Richard P. Aulie, "The Origin of the Idea of the Mammal-like Reptile." For a survey of changing interpretations of many aspects of reptile evolution, see Adrian J. Desmond, *The Hot-blooded Dinosaurs*.

58. Russell, *Form and Function*, pp. 288–95.

59. Ibid., chap. 15, and A. Willey, *Amphioxus and the Ancestry of the Vertebrates*.

60. Charles F. O'Brien, "*Eozoön canadense:* the Dawn Animal of Canada."

61. William Bateson, "The Ancestry of the Chordata," reprinted in *Scientific Papers,* 1: 1-31, and *Materials for the Study of Variation,* preface. Bateson had studied *Balanoglossus* for a clue to the origin of the chordates.

62. See Francis Galton, *Natural Inheritance* and *Hereditary Genius.* On Galton's life, see Karl Pearson, *The Life, Letters and Labours of Francis Galton.* On the significance of his work, see J. S. Wilkie, "Galton's Contributions to the Theory of Evolution"; R. G. Swinburne, "Galton's Law—Formulation and Development"; Ruth Schwartz Cowan, "Francis Galton's Contributions to Genetics"; and Robert De Marrais, "The Double-edged Effect of Sir Francis Galton." See also William B. Provine, *The Origins of Theoretical Population Genetics.*

63. Galton, *Natural Inheritance,* pp. 18-34. On pp. 27-30, Galton compares evolution to the rolling of a polygon with temporary positions of stability on each of its sides.

64. Galton, "Discontinuity in Evolution," p. 369.

65. On biometry see P. Froggat and N. C. Nevin, "The 'Law of Ancestral Heredity' and the Mendelian-Ancestrian Controversy"; B. J. Norton, "The Biometric Defense of Darwinism"; and Provine, *The Origins of Theoretical Population Genetics,* chaps. 2 and 3.

66. Karl Pearson, "Regression, Heredity and Panmixia" and "Mathematical Contributions to the Theory of Evolution: On the Law of Ancestral Heredity."

67. Beatrice Bateson, *William Bateson: Naturalist,* p. 10. On Weldon's career, see the obituary by Pearson.

68. See Weldon's first report to the Royal Society's committee: "Attempt to Measure the Death-rate due to the Selective Destruction of *Carcinus moenas*"; also "Remarks on Variation in Animals and Plants." See also his Presidential address to the Zoological Section of the British Association (1898) and "A First Study of Natural Selection in *Clausilia laminata.*"

69. Pearson's philosophy of science was expounded in his classic *Grammar of Science,* the second edition of which contained extra material on evolution. See B. J. Norton, "Biology and Philosophy: the Methodological Foundations of Biometry."

70. See Lankester, "Are Specific Characters Useful?" and Weldon's letter to *Nature* (1896).

71. See Pearson, "On the Fundamental Conceptions of Biology"; Weldon, "Mendel's Laws of Alternative Inheritance in Peas"; and Weldon's review of De Vries's book. On the debate, see A. G. Cock, "William Bateson, Mendelism, and Biometry," and Lindley Darden, "William Bateson and the Promise of Mendelism."

72. Weismann's early researches were collected in his *Studien zur Descendenztheorie,* translated as *Studies in the Theory of Descent.*

73. Translations of Weismann's papers are to be found in his *Essays upon Heredity:* The final statement of the theory, *Das Keimplasma,* was translated as *The Germ Plasm.* On the origins of Weismann's views, see Frederick B. Churchill, "August Weismann and a Break from Tradition."

74. See Weismann, *On Germinal Selection* and *The Evolution Theory,* chaps. 25 and 26.

75. "The All-Sufficiency of Natural Selection" is the title of the article by Weismann replying to Herbert Spencer's defense of Lamarckism.

76. See, for instance, the discussion of Brown-Séquard's experiments in Romanes,

Darwin and After Darwin, 2: chap. 4; and also Romanes's *An Examination of Weismannism.*

77. Butler, "The Deadlock in Darwinism," reprinted from the *Universal Review* for 1890 in Butler, *Essays on Life, Art and Science,* pp. 234–340; see p. 236. See also Marcus Hartog's introduction to the third edition of Butler's *Unconscious Memory.*

78. Lankester reprinted his "A Theory of Heredity" (1876) in his *Advancement of Science* (1890), adding a note stating that he now rejected the Lamarckian views there expressed in the light of Weismann's work. See p. 286.

79. Spencer, "The Inadequacy of Natural Selection," and also his *Factors of Organic Evolution.*

CHAPTER 3

1. The standard work on this episode is Charles C. Gillispie, *Genesis and Geology.* More recent studies include Neal C. Gillespie, *Charles Darwin and the Problem of Creation,* and Michael Ruse, *The Darwinian Revolution.*

2. This is the central theme of James R. Moore, *The Post-Darwinian Controversies.* See also Alvar Ellegård, *Darwin and the General Reader;* Frank Miller Turner, *Between Science and Religion;* and R. M. Young, "The Impact of Darwin on Conventional Thought."

3. See Bowler, "Darwinism and the Argument from Design."

4. On idealism and the law of parallelism, see E. S. Russell, *Form and Function,* and Stephen Gould, *Ontogeny and Phylogeny.*

5. On Agassiz, see Edward Lurie, *Louis Agassiz.* On the parallel with the fossil record, see Gould, *Ontongeny and Phylogeny,* and Bowler, *Fossils and Progress.*

6. See Owen, *On the Nature of Limbs.*

7. The 12th edition of *Vestiges,* in which Chambers's name is finally acknowledged, was a reprinting of the 11th of 1860. On *Vestiges,* see Milton Millhauser, *Just Before Darwin,* and M. J. S. Hodge, "The Universal Gestation of Nature." On Chambers's use of fossil evidence, see Bowler, *Fossils and Progress,* chap. 3.

8. Given the technological limitations of his time, Babbage designed an extremely sophisticated computer, or "calculating engine." The quotation from his *Ninth Bridgewater Treatise* in which he argued that God could have "programmed" miracles into the universe from the beginning is still there in the 11th edition of *Vestiges;* see pp. 140–42.

9. Mivart, *Genesis of Species,* pp. 109–10. We shall encounter this analogy again in Galton's *Natural Inheritance.* On Mivart's critique of Darwinism, see Jacob W. Gruber, *A Conscience in Conflict,* and Peter J. Vorzimmer, *Charles Darwin: The Years of Controversy.*

10. *Genesis of Species,* pp. 84–87. For more details of these arguments, see Bowler, "Darwinism and the Argument from Design."

11. The fifth edition of Argyll's *Reign of Law* carries a response to Wallace's "Creation by Law"; see pp. 393–97. Wallace's paper is reprinted in *Natural Selection and Tropical Nature,* pp. 141–66.

12. Argyll's definitions of law can be found in *The Reign of Law,* pp. 64–65. Perhaps

the clearest statement of Argyll's belief that all structures were predesigned into life from the beginning can be found in his later book *Organic Evolution Cross-Examined,* pp. 115-19.

13. *The Reign of Law,* pp. 33-35 and 205-9.

14. Carpenter, "The Argument from Design in the Organic World," reprinted from the *Modern Review* for 1884 in Carpenter, *Nature and Man,* pp. 409-63. See Roger Smith, "The Human Significance of Biology."

15. Cope, "On the Origin of Genera," pp. 243-44 and 269. See Bowler, "Edward Drinker Cope."

16. Owen, *On the Anatomy of the Vertebrates,* 3: 805-9. See Roy M. MacLeod, "Evolutionism and Richard Owen."

17. *The Reign of Law,* pp. 233-48. See also Argyll's paper "On Variety as an Aim in Nature."

18. Moore, *The Post-Darwinian Controversies,* pp. 269-80. See Gray's papers on design in his *Darwiniana* and the biography by A. Hunter Dupree.

19. Gray, *Darwiniana,* pp. 121-22.

20. See the conclusion of Darwin's *Variations of Animals and Plants under Domestication.*

21. *The Reign of Law,* p. 213, and more especially the note "Prophetic Germs" and the letters to Sir William Flower, Argyll, *Autobiography and Memoirs,* 2: 483.

22. Gray, *Darwiniana,* pp. 57-58. Gray explicitly refused to accept the Lamarckism of the American school and denied the role of any external factor in eliciting favorable variations; see Dupree, *Asa Gray,* p. 357.

23. Butler, *Evolution, Old and New,* p. 48.

24. Ibid., pp. 31-32.

25. For Lamarck, as for Chambers himself, the inheritance of acquired characters was only a secondary mechanism interfering with the regular ascent of life up the chain of being. On Lamarck's theory, see Richard W. Burkhardt, Jr., *The Spirit of System.*

26. See Chambers, *Vestiges,* 11th ed., p. lxiii, quoting earlier editions of the same work.

27. Ibid., pp. 160-61.

28. Mivart, *Genesis of Species,* p. 74.

29. Butler, *Evolution, Old and New,* pp. 342-43 and *Luck, or Cunning,* pp. 16-19.

30. Mivart, "On the Development of the Individual and of the Species," p. 472.

31. Carpenter, "On the Hereditary Transmission of Acquired Psychical Habits."

32. Argyll, "Organic Evolution" and "Acquired Characters and Congenital Variations."

33. Dewey, *The Influence of Darwin on Philosophy,* p. 12.

34. Henderson, *The Fitness of the Environment.*

35. The edition of Kropotkin's *Mutual Aid* cited has an introduction by Ashley Montagu, who supports this anti-Darwinian position; see also his *Darwin: Competition and Cooperation.*

36. Bather, "Pseudo-Biology."

37. Geddes and Thomson, *The Evolution of Sex,* pp. 279–81. Note, however, that in a later work the same authors did accept a rather more positive role for the development of altruism; see their *Evolution,* pp. 175–77 and 245–48.

38. Kropotkin, "The Direct Action of the Environment on Plants" and "Inheritance of Acquired Characters."

39. Bergson, *Creative Evolution,* pp. 76–84.

40. Ibid., pp. 86–119. Bergson refers to the works of Cope and Eimer for the evidence of orthogenesis.

41. Shaw's encounter with Bergson is described in Bertrand Russell, *Portraits from Memory,* p. 73.

CHAPTER 4

1. Gould, *Ontogeny and Phylogeny,* p. 423, n. 20. On Lamarck's own theory see, for instance, Burkhardt, *The Spirit of System.*

2. See Conway Zirkle, "The Early History of the Ideas of the Inheritance of Acquired Characters and Pangenesis."

3. Packard first used the term in the introduction to *The Standard Natural History* (1885). His biography of Lamarck in 1901 contained the first extensive translations of Lamarck's work into English. Lamarck's *Philosophie zoologique* was not translated entire until 1914, and appeared then with an introduction by Hugh Elliot admitting that Lamarckism had lost much of its influence.

4. Lankester, "Inheritance of Acquired Characters."

5. See, for instance, Delage and Goldsmith, *The Theories of Evolution,* p. 251.

6. Spencer, *The Factors of Organic Evolution,* p. 29.

7. Butler, "The Deadlock in Darwinism," *Essays,* p. 309. Butler also states that ten years earlier Lamarck's name was only mentioned as a byword for extravagance.

8. Kropotkin, "The Direct Action of the Environment on Plants," p. 58.

9. Haldane, *Possible Worlds,* p. 31.

10. Poulton, "Acquired Characters," following Lankester's note under the same title.

11. J. Arthur Thomson, *Heredity,* pp. 179–91.

12. Gould, *Ontogeny and Phylogeny,* chap. 4.

13. Steele, *Somatic Selection and Adaptive Evolution.*

14. Weismann's classic experiment is described in his "The Supposed Transmission of Mutilations," *Essays upon Heredity,* 1: 433–61.

15. Darwin's theory of pangenesis is described in the penultimate chapter of his *Variation of Animals and Plants under Domestication.* See Gerald L. Geison, "Darwin and Heredity."

16. Eiseley's *Darwin's Century* makes perhaps the most extravagant claims about the extent of Darwin's retreat into Lamarckism.

17. See, for instance, Spencer, *The Factors of Organic Evolution*, pp. 29–33.

18. Darwin, *Variations of Animals and Plants*, chap. 23.

19. Quoted in D. S. Jordan and V. Kellogg, *The Scientific Aspects of Luther Burbank's Work*, pp. 85–86. Opponents also conceded that the opinion of "practical men" was strongly in favor of the inheritance of acquired characters; see, for instance, Thomson, *Heredity*, pp. 193–94.

20. See, for instance, Elizabeth Lomax, "Infantile Syphilis as an Example of Nineteenth Century Belief in the Inheritance of Acquired Characteristics."

21. Brown-Séquard's first experiments were reported in 1859 and were already known in Britain, thanks to his announcement in the *Proceedings of the Royal Society* for 1860. See J. M. D. Olmsted, *Charles Édouard Brown-Séquard*, pp. 172–74.

22. See, for instance, William H. Montgomery, "Germany," in Glick, *The Comparative Reception of Darwinism*, p. 101.

23. Semper's *Natürlichen Existenzbedingungen der Thiere* (1880) was translated as *The Natural Conditions of Existence as they Affect Animal Life* (1881). See p. 17.

24. See Daniel Gasman, *The Scientific Origins of National Socialism*.

25. Haeckel, *History of Creation*, 1: 111–16.

26. Ibid., 1: 213–16.

27. An account of Haeckel's *Perigenesis der Plastidule* can be found in Lankester, "A Theory of Heredity," reprinted in his *Advancement of Science*.

28. Hering's *Gedächtniss als allgemeine Funktion der organisirten Materie* (Vienna, 1870) is translated as "On Memory as a Universal Function of Organized Matter" in Samuel Butler, *Unconscious Memory*, chap. 6.

29. Haeckel, *Riddle of the Universe*, p. 122.

30. Ibid., p. 296. On Haeckel's monism, see Gasman, *Scientific Origins of National Socialism*, and Niles Holt, "Ernst Haeckel's Monistic Religion."

31. Haeckel, *Riddle of the Universe*, pp. 132–33.

32. Reprinted in Spencer, *Essays*, 1: 381–87. Note that the first sentence of the original 1852 version reads "theory of Lamarck" rather than "theory of evolution."

33. Spencer, *Principles of Biology*, 1: 256.

34. Ibid., pp. 181–83 and 254–56.

35. Ibid., p. 272.

36. Ibid., pp. 250–51.

37. Ibid., pp. 405–6.

38. Lankester, "A Theory of Heredity" (see n. 27).

39. See, for instance, the preface to *The Factors of Organic Evolution*. On Spencer's social philosophy, see J. D. Y. Peel, *Herbert Spencer: The Evolution of a Sociologist*.

40. *The Factors of Organic Evolution* originally appeared as articles in the *Nineteenth Century* in 1886.

41. Spencer, "The Inadequacy of Natural Selection." Weismann's reply was "The All-Sufficiency of Natural Selection," and Spencer's response to this was "Professor Weismann's Theories." See Frederick B. Churchill, "The Weismann-Spencer Controversy."

42. Spencer, "The Inadequacy of Natural Selection," p. 446.

43. Ibid., pp. 156–59.

44. Ibid., p. 456.

45. Thomson, *Heredity*, p. 166.

46. See, for instance, Haeckei's *Evolution of Man*, 5th ed., p. 349.

47. See Butler, *Life and Habit*, pp. 1–2 and 41. On Butler's views, see Basil Willey, *Darwin and Butler;* and for background to his work, see H. F. Jones, *Samuel Butler: A Memoir.*

48. See Francis Darwin, "President's Address" (to the British Association, 1908), pp. 15–16n., and H. F. Jones, *Charles Darwin and Samuel Butler.* On Butler's influence in science, see also Marcus Hartog's introduction to the second edition of *Unconscious Memory.* Butler also rates a chapter in E. S. Russell's *Form and Function* (1916).

49. Butler was working with Ribot's *L'hérédité* and Carpenter's *Mental Evolution in Animals.*

50. Butler, *Life and Habit*, pp. 125–27; see also *Unconscious Memory*, p. 17.

51. *Life and Habit*, chap. 14; see also *Evolution, Old and New*, p. 80.

52. *Life and Habit*, pp. 202–3.

53. Most historians now agree that Butler greatly exaggerated the extent of Buffon's evolutionism. His own favorite evolutionist seems to have been Erasmus Darwin rather than Lamarck. On the quarrel with Charles Darwin, see Jones, *Samuel Butler: A Memoir*, app. C.

54. *Evolution, Old and New*, pp. 31–32. Butler also wrote to Mivart emphasizing that Lamarckism preserved design, and that it was better to internalize the Designer rather than to abandon Him altogether; see Jones, *Samuel Butler*, 1: 407.

55. Translation of Hering (n. 28).

56. Butler, *Notebooks*, p. 66.

57. *Unconscious Memory*, p. 178.

58. *Luck or Cunning*, pp. 266–67.

59. Butler, *Essays*, p. 308.

60. Details of the controversy raging in the pages of *Nature* may be found in the introduction to the translation of Weismann's *Essays on Heredity.*

61. I have already noted the objections listed in Thomson's *Heredity*, pp. 179–91. See also W. Platt Ball, *Are the Effects of Use and Disuse Inherited?*

62. Kropotkin, "The Direct Action of the Environment on Plants," p. 62n.

63. Thomson's *Heredity* attributes the phrase "interminable question" to W. K. Brooks. Brooks had certainly lost patience with Lamarckism (see his *Foundations of Zoology*, lecture 4), although I have not found the phrase used here.

64. This objection was repeated on numerous occasions. It appeared at least twice in the pages of *Nature:* in 1898 with Edward Fry's "What are Acquired Characters?" (which sparked the debate leading to Lankester's and Poulton's comparisons of Lamarck's two laws) and in 1908 with Archibald Reid's "The Inheritance of 'Acquired' Characters." J. Arthur Thomson criticized W. K. Brooks and George Sandeman for advocating this view—Brooks in *Foundations of Zoology* and Sandeman in

Problems of Biology—see Thomson, *Heredity*, pp. 177–78. Much later J. T. Cunningham rejected the same point, which had been advanced in E. T. Goodrich's *Living Organisms;* see Cunningham, *Modern Biology*, pp. 102–3.

65. T. H. Montgomery, *Analysis of Racial Descent in Animals*, p. 146. Montgomery believed that a changed environment acted dirctly on the germ plasm to produce nonadaptive variation.

66. John A. Ryder, "Proofs of the Effects of Habitual Use in the Modification of Animal Organisms." On Ryder's work, see Chap. 6.

67. Arthur Dendy, "Progressive Evolution and the Origin of Species," p. 395.

68. Ibid., p. 393.

69. Delage, *L'hérédité*, p. 861 (my translation).

70. Hans Driesch, *The Science and Philosophy of the Organism*, 1: 272. Although Driesch had commented on evolution before, this work (the Gifford Lectures for 1907 and 1908) contains his major statement on the topic.

71. Ibid., pp. 282–89, referring to August Pauly's *Darwinismus und Lamarckismus* (1905). Driesch saw Pauly's concept as central to Lamarckism, whereas in fact it bears more resemblance to the mechanism of "organic selection" discussed later in this chapter.

72. Félix Le Dantec, *The Nature and Origin of Life*, pp. 119–29.

73. Richard Semon, *Die Mneme*, for example pp. 380–84; *The Mneme*, pp. 275–77.

74. See, for instance, Thomson, *Heredity*, p. 172, and Kropotkin, "The Direct Action of the Environment on Plants," p. 77.

75. See, for instance, Thomson, *Heredity*, p. 172. Thomson later tried to promote a very optimistic image of Darwinism; see *The Gospel of Evolution*.

76. C. Lloyd Morgan, *Habit and Instinct*, p. 312. The relevant passage is also quoted in app. A of James Mark Baldwin's *Development and Evolution*, which describes the origin of the idea. The Baldwin effect was given prominence by Sir Alistair Hardy in *The Living Stream*, but for a less positive evaluation see George Gaylord Simpson, "The Baldwin Effect."

77. Kropotkin, "The Direct Action of the Environment on Plants," p. 77n.

78. The American school is discussed in Chap. 6. Orr's book is *A Theory of Development and Heredity*.

79. Hartog, "The Fundamental Principles of Heredity." On Hartog's efforts to get this article published, see his introduction to Butler's *Unconscious Memory*, p. xxiv.

80. Hartog, "Fundamental Principles of Heredity," p. 316.

81. Hartog, introduction to Butler's *Unconscious Memory*, pp. xxxiv–xxxv.

82. See H. F. Jones, *Charles Darwin and Samuel Butler*.

83. Francis Darwin, "President's Address," p. 14.

84. Ibid., p. 4.

85. Eugenio Rignano, *Upon the Inheritance of Acquired Characters*.

86. Ibid., pp. 351–52.

87. This statement occurs in the English translation, *The Mneme*, p. 14, from a section that appears to be an addition to the third German edition (the translation was prepared under Semon's supervision).

88. Semon, *Die Mneme*, p. 384; *The Mneme*, p. 277.

89. *Die Mneme*, pp. 403–4n. Although omitted from the English version, the note was translated in Hartog's introduction to Butler's *Unconscious Memory*, pp. xxxiv-xxxv, where *PsychoLamarckisten* is rendered as "neo-Lamarckians."

90. *Die Mneme*, p. 386; *The Mneme*, pp. 278–79.

91. *Die Mneme*, chap. 13.

92. George Henslow, *Origin of Floral Structures*, pp. vi–vii. Henslow had already published works on the problems of evolution and religion.

93. Ibid., pp. 331–33.

94. Reported in Henslow, "Origin of Plant Structures," and the book under the same title.

95. Henslow, *Origin of Plant Structures*, p. 9.

96. Ibid., p. viii.

97. A. R. Wallace, "The Rev. George Henslow on Natural Selection."

98. Henslow, "The Origin of Species without the Aid of Natural Selection."

99. Henslow, "Does Natural Selection Play any Part in the Origin of Species?" and "Scientific Proofs vs. '*A Priori*' Assumptions." See the critical article by J. Lionel Taylor, "The Study of Variations."

100. Romanes, *Darwin and After Darwin*, 2: 18–19.

101. Kropotkin, "The Direct Action of the Environment on Plants," p. 76.

102. Ibid., pp. 63–66. Several works by Bonnier are listed in the Bibliography.

103. Herbert Spencer Jennings, *Behavior of the Lower Organisms*. For Jennings's own views on the evolution of behavior, which were not Lamarckian, see his chap. 19.

104. Weismann, *Studies in the Theory of Descent*, 1: 108. Standfuss's experiments were summarized in English in his "Synopsis of Experiments on Hybridization and Temperature made with Lepidoptera."

105. Weismann, *Germ Plasm*, pp. 402–3.

106. See, for example, H. M. Vernon, *Variation in Animals and Plants*, pp. 357–58 and Kellogg, *Darwinism Today*, p. 319, which discusses the case of nonadaptive variation in beetles reported in Kellogg's "Is there Determinate Variation?"

107. See further material in this chapter and also Bowler, "Theodor Eimer and Orthogenesis."

108. Romanes, *Darwin and After Darwin*, 2: 103–22.

109. Olmsted, *Brown-Séquard*, pp. 190–95.

110. Surveys of the early twentieth-century Lamarckian experiments can be found in Kammerer, *Inheritance of Acquired Characteristics*, especially chap. 38; G. C. Robson and O. W. Richards, *The Variation of Animals in Nature*, especially pp. 35–42; and Phillip Fothergill, *Historical Aspects of Organic Evolution*, chap. 9. On the effects of alcohol, see Charles R. Stockard and Dorothy M. Craig, "An Experimental Study of the Influence of Alcohol on the Germ Cells"; and on the effects of rotation, J. A. Detlefsen, "Are the Effects of Long-Continued Rotation in Rats Inherited?" V. Kellogg and R. G. Bell also reported the inheritance of effects induced by starvation; see "Variations Induced in Larval, Pupal and Imaginal Stages of *Bombyx mori.*"

111. Walter Kidd, *Use-Inheritance.*

112. Kidd, *Initiative in Evolution.* Despite its title, this work confines itself to some general anti-Darwinian remarks and a more extensive exposition of the argument based on direction of hair.

113. Joseph T. Cunningham, "Blind Animals in Caves."

114. Cunningham, "The New Darwinism." For Cunningham's later views on the social implications of Lamarckism, see the conclusion of his *Modern Biology.*

115. See, for instance, Cunningham, *Sexual Dimorphism in the Animal Kingdom.*

116. Cunningham, "The Utility of Specific Characters," written in response to Wallace's "The Problem of Utility."

117. Cunningham, "The Problem of Variation," pp. 283-84.

118. Cunningham and C. A. MacMunn, "On the Coloration of the Skins of Fishes," and Cunningham, "The Problem of Variation." The first paper was communicated to the Royal Society by Lankester, which perhaps explains Cunningham's reluctance to engage in Lamarckian theorizing.

119. Cunningham, "The Origin and Evolution of Flatfishes."

120. Cunningham, *Sexual Dimorphism in the Animal Kingdom.*

121. Cunningham, *Hormones and Heredity.*

122. Theodor Eimer, *Die Entstehung der Arten,* translated by Cunningham as *Organic Evolution.* Cunningham's preface gives a good summary of his own views.

123. Eimer, *Organic Evolution,* p. 381.

124. Ibid., p. 434.

125. Rudolph Virchow, "Descendenz und Pathologie." Virchow's support for Lamarckism is described in detail by F. B. Churchill, "Rudolph Virchow and the Pathologists's Criteria for the Inheritance of Acquired Characters."

126. Oscar Hertwig, *Die Zelle und Gewebe,* 2: 241-43. On Hertwig's opposition to Weismann in cytology, see F. B. Churchill, "Hertwig, Weismann, and the Meaning of Reduction Division."

127. Hertwig, *Die Zelle und Gewebe,* 2: 273-76.

128. Hertwig, *Das Werden der Organismen.*

129. Paul Kammerer, "Breeding Experiments on the Inheritance of Acquired Characters," p. 637.

130. See, for instance, Kammerer, "Adaptation and Inheritance in the Light of Modern Experimental Investigation." This is a translation of a popular account of his work given by Kammerer to a scientific society in Berlin in 1910.

131. The most detailed account of the experiments is in Kammerer, "Die Nachkommen der nicht Brutpflegenden *Alytes obstetricans.*" Brief accounts appear in both English-language papers by Kammerer and in his 1924 book. For modern interpretations of his results, see, for instance, C. H. Waddington, *The Evolution of an Evolutionist,* pp. 177-81, and Lester R. Aronson, "The Case of *The Case of the Midwife Toad.*" In addition to Koestler's book, there are impressions of Kammerer in H. G. Cannon, *Lamarck and Modern Genetics,* and in A. M. Werfel, *And the Bridge is Love.*

132. Kammerer, "Adaptation and Inheritance," p. 434. More specifically, in the

third generation the males developed rough swellings on their forelimbs and in the fourth generation these became blackened; see "Die Nachkommen der nicht Brutpflegenden *Alytes obstetricans*," p. 535.

133. Kammerer, "Adaptation and Inheritance," p. 435.

134. Kammerer's most important paper is "Das Farbkleid des Feuersalamanders, *Salamandra maculosa Laurenti*, in seiner Abhängigkeit von der Umwelt." Again, the experiments are described in all of Kammerer's English-language works and in Koestler's book.

135. Kammerer, "Adaptation and Inheritance," p. 439.

136. Kammerer, *The Inheritance of Acquired Characteristics*, pp. 251-52.

137. Kammerer, "Breeding Experiments," p. 638, and *The Inheritance of Acquired Characteristics*, chap. 18.

138. See Koestler, *Case of the Midwife Toad*, which lists all the contributions to the debate. Note, however, that Koestler misrepresents the situation by implying that Bateson was opposing Kammerer from a Darwinian position—in fact, he was an equally implacable critic of the selection theory.

139. William Bateson, *Problems of Genetics*, p. 199.

140. Kammerer, "Breeding Experiments," p. 639, and *The Inheritance of Acquired Characteristics*, p. 64.

141. Kammerer discusses this objection in *The Inheritance of Acquired Characteristics*, pp. 47 and 72, and mentions an interview with Keith published in the *Daily Express* for 2 May 1923. Kammerer points out that such an argument would make the inheritance of acquired characters unprovable.

142. Examples of the press coverage are given in Koestler, *Case of the Midwife Toad*.

143. Kammerer, *Inheritance of Acquired Characteristics*, chap. 43.

144. On Haeckel's contributions to race theory, see Daniel Gasman, *The Scientific Origins of National Socialism;* and on the American school, John S. Haller, *Outcasts from Evolution.*

145. E. W. MacBride, *Introduction to the Study of Heredity*, chap. 9. MacBride was professor of zoology at Imperial College, London. For a brief outline of his work, see Richard W. Burkhardt, Jr., "Lamarckism in Britain and the United States."

146. MacBride, *Introduction*, pp. 60-61 and 83. This book was a contribution to the "Home University Library," although it was subsequently replaced by a more orthodox text.

147. Ibid., p. 147; and on the gene theory, chap. 2.

148. MacBride's popular little text *Evolution* still contains a chapter proclaiming the importance of recapitulation. On the decline of the theory see Gould, *Ontogeny and Phylogeny*, and Frederick B. Churchill, "The Modern Evolutionary Synthesis and the Biogenetic Law."

149. William MacDougall, "An Experiment for the Testing of the Hypothesis of Lamarck," p. 303.

150. Ibid. Accounts of the experiment can be found in Fothergill, *Historical Aspects of Organic Evolution*, pp. 261-71, and D. R. Oldroyd, *Darwinian Impacts*, chap. 13.

151. MacDougall, "An Experiment," p. 268.

152. For details of the critique, see Oldroyd, *Darwinian Impacts.*

153. Kammerer, *Inheritance of Acquired Characteristics,* p. 19.

154. The complete proceedings of the symposium may be found in *Proceedings of the American Philosophical Society 42* (1923): 270-325. This includes the report by Detlefsen cited above in n. 110 and another by Stockard, "Experimental Modification of the Germ Plasm."

155. J. W. Heslop Harrison and F. C. Garrett, "The Induction of Melanism in the Lepidoptera." This experiment is described by Alistair Hardy, *The Living Stream,* p. 119, where it is pointed out that J. B. S. Haldane criticized it on the grounds that a recessive melanic gene was already present in the original population.

156. See Kammerer, *Inheritance of Acquired Characteristics,* p. 197, and MacBride, *Introduction to Heredity,* chap. 8. MacBride argues that genetic damage caused by poisons is the chief source of mutation.

157. Heslop Harrison, "Experiments on the Egg-laying Instincts of the Saw-fly." These experiments were criticized by Haldane in "The Hereditary Transmission of Acquired Characters" and defended by MacBride in "The Inheritance of Acquired Characters." Julian Huxley attributed the results to organic selection; see *Evolution: the Modern Synthesis,* p. 304. See Fothergill, *Historical Aspects of Organic Evolution,* pp. 258-61.

158. See especially the contributions by Stephen Gould, Bernhard Rensch, and Viktor Hamburger in Mayr and Provine, *The Evolutionary Synthesis.*

159. Rensch, *Das Prinzip geographischer Rassenkreise;* see also Rensch, *Evolution Above the Species Level,* p. 188. Allen's law is named after the American naturalist Joel Allen, discussed in Chap. 6.

160. G. C. Robson, *The Species Problem,* pp. 158-72, 255-57.

161. Robson and Richards, *The Variation of Animals in Nature,* chap. 2; and for a conclusion on selection, see p. 316.

162. Ibid., pp. 343-44.

163. George Bernard Shaw, *Preface to Back to Methuselah: A Metabiological Pentateuch;* see *The Bodley Head Bernard Shaw,* 5: 303.

164. See Lloyd Morgan, *Emergent Evolution;* Samuel Alexander, *Space, Time and Deity;* Roy Wood Sellars, *Evolutionary Naturalism;* and Alfred North Whitehead, *Process and Reality.* See also Dorothy Emmet, *Whitehead's Philosophy of Organism.*

165. Jan Christiaan Smuts, *Holism and Evolution,* chap. 8. Smuts was the famous South African soldier and statesman.

166. See, for instance, Zhores Medvedev, *The Rise and Fall of T. D. Lysenko,* and D. Joravsky, *The Lysenko Affair.*

CHAPTER 5

1. For biographical information, see J. M. D. Olmsted, *Charles-Eduard Brown-Séquard.*

2. The most extensive study of the French reaction to Darwinism is Yvette Conry, *L'introduction du Darwinisme en France au XIXe siècle.* See also John Farley,

"The Initial Reaction of French Biologists to Darwin's *Origin of Species*"; and Robert E. Stebbins, "France," in Glick, *The Comparative Reception of Darwinism*, pp. 117-63. On the situation at the turn of the century, see Camille Limoges, "Natural Selection, Phagocytosis, and Preadaptation: Lucien Cuénot, 1886-1901." On twentieth-century French evolutionism, see Ernest Boesiger, "Evolutionary Biology in France at the Time of the Evolutionary Synthesis," and Limoges, "A Second Glance at Evolutionary Biology in France."

3. Cuénot, *L'évolution biologique*, pp. v–vi.

4. Farley, "The Initial Reaction of French Biologists."

5. See Harry W. Paul, *The Edge of Contingency: French Catholic Reaction to Scientific Change from Darwin to Duhem.*

6. The title was *De l'origine des espèces, ou des lois du progrès dans les êtres organisés.*

7. Yves Delage, *L'hérédité et les grands problèmes de la biologie générale*, p. 322.

8. Ibid., p. 861. My translation.

9. Edmond Perrier, "Le transformisme et les sciences physiques."

10. Perrier and Charles Gravier, "La tachygénèse."

11. Perrier, *La philosophie zoologique avant Darwin*, preface. See chapter 8 on Lamarck.

12. Alfred Giard, "l'évolution des êtres organisés."

13. Giard, "Les facteurs de l'évolution."

14. Ibid., p. 646.

15. Giard, "Le principe de Lamarck et l'hérédité des modifications somatiques."

16. Ibid., p. 713.

17. Conry, *L'introduction du Darwinisme*, p. 369.

18. E.g., Ernst Mayr, "Some Thoughts on the History of the Evolutionary Synthesis," p. 15.

19. Félix Le Dantec, *The Nature and Origin of Life*, pp. 115–16. Le Dantec's views are described in Delage and Goldsmith, *Theories of Evolution*, pp. 261-70.

20. Le Dantec, *Lamarckiens et Darwiniens*, pp. 45–65.

21. Ibid., p. 28. Le Dantec conceded that this was not a necessary relationship, but argued that the evidence was in its favor.

22. Ibid., part 3, pp. 100-153.

23. Ibid., pp. 13-17.

24. Steele's *Somatic Selection and Adaptive Evolution* explains the working of the immune system by postulating rapid mutation in the production of antibodies.

25. Le Dantec, *Nature and Origin of Life*, p. 129. Robert Duncan's preface to this English edition explains that Le Dantec was asked to prepare a statement of his materialist view of life for the English-speaking audience.

26. Ibid., p. 237.

27. Elie Metchnikoff, "La lutte pour l'existence entre les diverses parties de l'organisme."

28. For details of this part of Cuénot's career, see Limoges, "Natural Selection, Phagocytosis, and Preadaptation."

29. Cuénot, L'évolution biologique.

30. A. Delcourt and Emile Guyenot, "Genetique et milieu."

31. Maurice Caullery, Le problème de l'évolution.

32. Jean Rostand, Les chromosomes, artisans de l'hérédité et du sexe.

33. Cuénot, La genèse des espèces animales, p. 439.

34. Bergson, Creative Evolution, e.g., p. 72. I have already noted Bergson's philosophy as a reformulation of the old argument from design in Chap. 3 above.

35. Jacques Monod recalls the influence of Bergson during his youth in Chance and Necessity, p. 34.

CHAPTER 6

1. Edward J. Pfeifer, "The Genesis of American Neo-Lamarckism."

2. On the role played by the recapitulation theory in American neo-Lamarckism, see Gould, Ontogeny and Phylogeny, pp. 85-100.

3. See Bowler, "Theodor Eimer and Orthogenesis," and chap. 7 below.

4. G. G. Simpson, The Major Features of Evolution, p. 266.

5. On Agassiz's influence, see Edward Lurie, Louis Agassiz; Pfeifer, "United States," in Glick, ed., The Comparative Reception of Darwinism; Gould, Ontogeny and Phylogeny; and Ralph W. Dexter, "The Impact of Evolutionary Theories on the Salem Group of Agassiz Zoologists." For a more general discussion of the continuing influence of morphological studies, see Ronald Rainger, "The Contribution of the Morphological Tradition."

6. See Darwin's correspondence with Hyatt, Life and Letters of Darwin, 3: 154, and More Letters, 1: 338-48; also his letter to E. S. Morse, Life and Letters, 3: 233.

7. See, for instance, Robert Plate, The Dinosaur Hunters; and for biographical material on Cope, see Theodor Gill, "Edward Drinker Cope, Naturalist," and Henry Fairfield Osborn, Cope: Master Naturalist.

8. Also reprinted in Cope, The Origin of the Fittest, pp. 41-123. For a more detailed analysis of this paper, see Bowler, "Edward Drinker Cope and the Changing Structure of Evolutionary Theory."

9. Cope, "Origin of Genera," p. 269.

10. Cope confessed that he had not read Lamarck when he proposed the "law of use and effort" in his "On Catagenesis," Origin of the Fittest, p. 423. The quotation referring to Spencer in Osborn, Cope: Master Naturalist, p. 529, is given as though from the brief 1871 paper, "The Laws of Organic Development," although in fact it is from Cope's own discussion of that paper in the introduction to Origin of the Fittest, pp. viii-ix.

11. Cope, The Primary Factors of Organic Evolution, pp. 45-58, quoting Joel A. Allen, "The Influence of Physical Conditions in the Genesis of Species."

12. Cope, Primary Factors, pp. 29-41.

13. Cope, "The Origin of the Specialized Teeth of the Carnivora" (1879) and "On the Trituberculate Type of Molar Tooth in the Mammalia" (1883), reprinted in *Origin of the Fittest*, pp. 363–67 and 359–62.

14. Cope, "The Method of Creation of Organic Forms," pp. 250–51.

15. Cope, *Primary Factors*, pp. 146–50.

16. Ibid., p. 24.

17. Statement prefaced to the first issue of the *American Naturalist* 1 (March 1867): 2.

18. Cope, "On Archaesthetism," reprinted in *Origin of the Fittest*, pp. 405–21. On Cope's metaphysical views, see Herbert W. Schneider, *A History of American Philosophy*, pp. 359–63, and James R. Moore, *The Post-Darwinian Controversies*, pp. 148–51.

19. Later editions of LeConte's book were retitled *Evolution: Its Nature, Its Evidences, and Its Relation to Religious Thought*. On his social theory, see Lester G. Stephens, "Joseph LeConte's Evolutionary Idealism." More generally, see George Stocking, "Lamarckianism in American Social Thought," and William Coleman, "Science and Symbol in the Turner Frontier Hypothesis."

20. LeConte, *Evolution*, pp. 81–90.

21. The social implications of the recapitulation theory are illustrated in Gould, *Ontogeny and Phylogeny*, chap. 5.

22. Cope's numerous articles on race theory are discussed in John S. Haller, *Outcasts from Evolution*, pp. 192–202.

23. See Edward S. Morse, "Memorial of Professor Alpheus Hyatt," p. 417, and W. K. Brooks, "Biographical Memoir of Alpheus Hyatt," p. 314.

24. Hyatt, "On the Parallelism between the Different Stages of Life," p. 193.

25. Hyatt, *Genesis of the Arietidae*, p. 26.

26. Hyatt's later works are listed in the Bibliography below. The best short summaries of his ideas are "Evolution of the Cephalopoda" and "Cycle of Life in the Individual (Ontogeny) and in the Evolution of the Group (Phylogeny)."

27. Hyatt, "Evolution of the Cephalopoda," pp. 147–48.

28. Ibid., p. 125.

29. Hyatt, "The Influence of Woman in the Evolution of the Human Race."

30. See, for instance, Hyatt, "Bioplastology and Related Branches of Biologic Research."

31. For biographical information, see W. B. Scott, *Memoirs of a Paleontologist;* G. G. Simpson, "Biographical Memoir of William Berryman Scott;" and William K. Gregory, "Biographical Memoir of Henry Fairfield Osborn."

32. Scott, "On the Osteology of *Mesohippus* and *Leptomeryx*," pp. 363–64 and 370–71.

33. Ibid., p. 386.

34. Scott, "On Variations and Mutations." This article was written in reply to William Bateson's support for discontinuous evolution in his *Materials for the Study of Variation*.

35. Scott, *The Theory of Evolution* and *The History of Land Mammals*.

36. Osborn, "Paleontological Evidence for the Transmission of Acquired Characters" and "Are Acquired Characters Inherited?"

37. Osborn, "The Hereditary Mechanism and the Search for the Unknown Factors in Evolution."

38. Osborn's "A Mode of Evolution Requiring neither Natural Selection nor the Inheritance of Acquired Characters (Organic Selection)" is reprinted from the *Transactions of the New York Academy of Science* of 1896 in James Mark Baldwin, *Development and Evolution*, pp. 335–52. On the Baldwin effect, see above, Chap. 4.

39. Baldwin, *Development and Evolution*, pp. 160–64.

40. Osborn, "The Limits of Organic Selection."

41. For Osborn's views on human evolution, see his *Man Rises to Parnassus*. He was elected president of the second International Congress of Eugenics in 1927.

42. See, for instance, Osborn's *Origin and Evolution of Life*.

43. Packard's early interest in Lamarck is recorded in his diary, which is quoted at length in T. D. A. Cockerell, "Biographical Memoir of Alpheus Spring Packard." See also Dexter, "The Development of A. S. Packard, Jr., as a Naturalist and Entomologist."

44. See the account of the origins of the American school in Packard, *Lamarck: The Founder of Evolution*, chap. 20.

45. Packard, "On the Development of *Limulus polyphemus*," p. 198. The original article is "on the Embryology of *Limulus polyphemus*."

46. Packard and F. W. Putnam, *The Mammoth Cave and its Inhabitants*.

47. Packard, *Lamarck: The Founder of Evolution*, p. 157.

48. Packard, "On the Inheritance of Acquired Characters in Animals with a Complete Metamorphosis."

49. Packard, "Hints on the Evolution of the Bristles, Spines and Tubercules of Certain Caterpillars."

50. Packard, "The Cave Fauna of North America." See also "On the Origin of the Subterranean Fauna of North America."

51. Packard, "On the Inheritance of Acquired Characters in Animals with a Complete Metamorphosis," pp. 367–68n., referring to his introduction to the *Standard Natural History* (1885).

52. On natural theology, see Packard, "The Law of Evolution." For a brief comment on Lamarckism and civilization, see "On the Inheritance of Acquired Characters in Animals with a Complete Metamorphosis," p. 340.

53. Dexter calls Packard the "reconciler of neo-Lamarckism and Darwinism" on the basis of the sympathetic attitude toward Darwin expressed in the biography of Lamarck; see Dexter, "The Impact of Evolutionary Theories."

54. Packard, "On the Inheritance of Acquired Characters in Animals with a Complete Metamorphosis," p. 338.

55. Packard, *Lamarck: The Founder of Evolution*, pp. 404–5.

56. Nathaniel S. Shaler, "Lateral Symmetry in Brachiopoda." For Shaler's later views see, for instance, his *Interpretation of Nature*. By the end of the century he was promoting a loose mixture of optimistic Lamarckism and hereditarian race theory.

57. Allen, "The Influence of Physical Conditions in the Genesis of Species." Part of this article is reprinted as an example of Lamarckism in George Daniels, *Darwinism Comes to America*, pp. 77-84. For Allen's views on the adaptive value of the effects he studied, see "Remarks on the Geographical Variation of Mammals and Birds" and "Geographical Variation in North American Birds."

58. John A Ryder, "On the Mechanical Genesis of Tooth Forms." He later claimed that everyone except Cope had ignored the article; see "Proofs of the Effects of Habitual Use in the Modification of Animal Organisms," p. 549.

59. Ryder, "Proofs of the Effects of Habitual Use," for example, p. 544.

60. Ryder, "On Like Mechanical (Structural) Conditions Producing Like Morphological Effects," p. 159.

61. Ibid.

62. Ryder, "On the Origin of Bilateral Symmetry and the Numerous Segments of the Soft Rays of Fishes," and "Proofs of the Effects of Habitual Use." He tried to distinguish between the effects of static and dynamic forces; see "Energy as a Factor in Organic Evolution" and "Dynamics in Evolution."

63. Ryder, "The Inheritance of Modifications due to Disturbances of the Early Stages of Development."

64. Ryder, "A Physiological Hypothesis of Heredity and Variation" and "A Dynamical Hypothesis of Inheritance."

65. Ryder, "Proofs of the Effects of Habitual Use," p. 549.

66. See, for instance, Henry Orr, *A Theory of Development and Heredity*.

67. Charles R. Stockard and Dorothy M. Craig, "An Experimental Study of the Influence of Alcohol on the Germ Cells," and Stockard and George Pananicolaon, "A Further Analysis of the Hereditary Transmission of Degeneracy and Deformation." In the 1923 symposium see Stockard, "Experimental Modification of the Germ Plasm," and J. A. Detlefsen, "Are the Effects of Long-Continued Rotation in Rats Inherited?"

68. See Hamilton Cravens, *The Triumph of Evolution*.

CHAPTER 7

1. Ludwig Plate, *Über Bedeutung des Darwin'schen Selectionsprinzips*, p. 187. Julian Huxley's *Evolution: The Modern Synthesis* notes the importance of this concept and attributes it to Plate (p. 500), but mentions only a later work.

2. Oswald Spengler's *Untergang des Abendlandes* was first published in 1918.

3. See Stephen Gould, "Dollo on Dollo's Law."

4. Dollo collaborated in the production of a general work criticizing the idea of degeneration back to a primitive state in either biological or social evolution; see Jean Demoor et al., *Evolution by Atrophy*, preface.

5. This was the essence of the debate over the utility of specific characters. For a discussion see Chap. 1.

6. Carl Nägeli, *Entstehung des Begriffs der naturhistorischen Art*. For Darwin's admission that nonadaptive characters were his greatest difficulty, see *Descent of Man*, p. 61.

7. Nägeli, *A Mechanico-Physiological Theory of Organic Evolution*, p. 45.

8. Ibid., p. 30.

9. Ibid., p. 36.

10. Ibid., p. 8.

11. On orthogenesis, see Wilhelm Haacke, *Gestaltung und Vererbung*, p. 31. Haacke's views on the nature of the germ plasm are discussed in Delage, *L'hérédité*, pp. 470–81.

12. Theodor Eimer, *On Orthogenesis*, pp. 15n. and 20–22; *Orthogenesis der Schmetterlinge*, pp. 8–9n. and 13–16.

13. See, for instance, Venon Kellogg, *Darwinism Today*, pp. 282–85, which presents a mechanistic image of Eimer's theory and plays down the unifying effect of his trends.

14. Eimer, *Zoologische Studien auf Capri*, vol. 2, *Lacerta muralis caerula*. For a description of this work, see Bowler, "Theodor Eimer and Orthogenesis," pp. 46–47.

15. On the Lamarckism of Eimer's *Organic Evolution*, see above, Chap. 4. For an evaluation of the Eimer-Weismann debate by one of their contemporaries, see Bertram Windle's review of *Organic Evolution*.

16. This is Eimer, *On Orthogenesis and the Impotence of Natural Selection in Species-Formation*.

17. Ibid., p. 10.

18. Ibid., p. 52; *Orthogenesis der Schmetterlinge*, p. 38.

19. Ibid., p. 22; *Orthogenesis der Schmetterlinge*, p. 13. See also the passage translated in Kellogg, *Darwinism Today*, p. 285.

20. Eimer, *On Orthogenesis*, p. 36; *Orthogenesis der Schmetterlinge*, pp. 26–27. Standfuss's experiments are mentioned above in Chap. 4.

21. *On Orthogenesis*, p. 22; *Orthogenesis der Schmetterlinge*, p. 13.

22. *On Orthogenesis*, p. 21; *Orthogenesis der Schmetterlinge*, p. 14.

23. Eimer, *Organic Evolution*, pp. 409–10 and 433; *Entstehung der Arten*, pp. 440–42 and 460.

24. Weismann, *The Germ Plasm*, pp. 402–3. On Weismann's response to orthogenesis, see Bowler, "Theodor Eimer and Orthogenesis," pp. 53–58.

25. Weismann, "On Germinal Selection."

26. Weismann, *The Evolution Theory*, 2: 132–34 and 358.

27. Cunningham, "The Origin and Evolution of Flatfishes," p. 239.

28. Glenn L. Jepsen, "Selection, 'Orthogenesis,' and the Fossil Record," p. 490. This article contains a full bibliography on all aspects of orthogenesis.

29. This point is made most clearly in the case of invertebrate shell structures and is even more explicit in the later edition, where Thompson refers to work done by Arthur Dendy in the 1920s; see *On Growth and Form*, new ed., pp. 693–94. For more detailed descriptions of Thompson's views, see the biography by Ruth D'Arcy Thompson, *D'Arcy Wentworth Thompson: The Scholar Naturalist*, and Stephen Gould, "D'Arcy Thompson and the Science of Form."

30. Thompson, *On Growth and Form* (1917), p. 727.

31. The subtitle of *Nomogenesis* is *Evolution Determined by Law.* The laws themselves are outlined on pp. 154-55. J. B. S. Haldane called *Nomogenesis* "by far the best anti-Darwinian book of this century"; see *The Causes of Evolution*, p. 12.

32. Berg, *Nomogenesis*, e.g., pp. 115-16.

33. Ibid., p. 404.

34. Ibid., p. 114.

35. Ibid., foreword by Dobzhansky, pp. x-xi.

36. Bashford Dean, "Evolution in a Determinate Line, as Illustrated by the Egg-cases of Chimaeroid Fishes."

37. L. J. Henderson, "Orthogenesis from the Standpoint of the Biochemist."

38. Charles Otis Whitman, *Orthogenetic Evolution in Pigeons*, p. 11. For more details on Whitman's observations, see Bowler, "Theodor Eimer and Orthogenesis," pp. 67-69.

39. Ernst Mayr, *Dictionary of Scientific Biography*, s.v. "Whitman, Charles Otis."

40. For Whitman's general views on recapitulation, see *Orthogenetic Evolution in Pigeons*, chap. 10.

41. Wilhelm Waagen, "Die Formenreihe des *Ammonites subradius.*" The definition of mutation is on p. 186.

42. Waagen, *Jurassic Fauna of Kutch*, p. 242.

43. Alpheus Hyatt, "On the Parallelism between the Different Stages of Life" See above, Chap. 6.

44. Hyatt, "Lost Characteristics."

45. For a concise account of the process, see Hyatt, "Evolution of the Cephalopoda."

46. For a brief discussion of the work of these paleontologists, see Ronald Rainger, "The Continuation of the Morphological Tradition," pp. 152-57. For Beecher's views on the development of spines, see his *Studies in Evolution*.

47. The *Gryphaea* case is discussed by Lang (see below) and in H. H. Swinnerton's popular textbook, *Outlines of Palaeontology*, p. 222.

48. W. D. Lang, *Catalogue of the Fossil Bryozoa*, p. viii. See also Lang's popular lecture, "Evolution: A Resultant," which makes a general case for orthogenesis based also on the overcoiling of *Gryphaea* and—surprisingly—Eimer's butterflies.

49. Lang, *Catalogue of the Fossil Bryozoa*, p. vii.

50. Ibid., p. iv.

51. Ibid., p. xix. On Bateson's views, see below, Chap. 8.

52. Ibid., p. xviii.

53. Jepsen implies that it was the biochemists, not the paleontologists, who made the strongest effort to develop a theory of orthogenesis; see "Selection, 'Orthogenesis,' and the Fossil Record," p. 490.

54. A. Franklin Shull, "Weismann and Haeckel: One Hundred Years," p. 448.

55. Ludwig Döderlein, "Phylogenetische Betrachtungen"; Ernst Koken, *Palaeontol-*

ogie und Descendenzlehre; F. B. Loomis, "Momentum in Variation"; and Charles Depéret, *Les transformations du monde animal.*

56. A detailed discussion of the case of the Irish elk is given by Stephen Gould in "The Origin and Function of 'Bizarre' Structures."

57. Depéret, *Les transformations du monde animal,* p. 249. The translation is that of E. S. Russell, *Form and Function,* p. 361. See the full English translation, *The Transformations of the Animal World,* p. 243.

58. Arthur Dendy, "Momentum in Evolution." On Dendy's Lamarckism see above, Chap. 4.

59. D. M. S. Watson, "Croonian Lecture," p. 194. Another example of a supposedly nonadaptive trend is given by L. T. Hogben in "The Progessive Reduction of the Jugal in the Mammalia."

60. Nopsca, "Notes on Stegocephalia and Amphibia."

61. R. S. Lull, "Dinosaurian Climatic Response," p. 276. Lull was Sterling Professor of Paleontology at Yale.

62. Arthur Smith Woodward, "President's Address," p. 468. Woodward is better known for his role in the "Piltdown man" affair.

63. Loomis, "Momentum in Variation," p. 839.

64. Maynard Metcalf, "Adaptation through Natural Selection and Orthogenesis."

65. See W. D. Matthew, "The Evolution of the Horse," which is cited as pro-orthogenesis by Julian Huxley in *Evolution: The Modern Synthesis,* p. 498 (although in fact Matthew allows for a certain amount of branching in the process). For Matthew's opinion on the saber-toothed tiger, see "The Phylogeny of the Felidae," pp. 305-7.

66. W. B. Scott, *History of Land Mammals in the Western Hemisphere,* p. 532.

67. Scott, "On Variations and Mutations," p. 372. quoting his own "On the Osteology of Mesohippus," p. 388.

68. H. F. Osborn, "The Continuous Origin of Certain Unit Characters," pp. 187-88.

69. Some of Osborn's works establishing the nature of rectigradations are *The Evolution of Mammalian Molar Teeth;* "The Four Inseparable Factors of Evolution"; "The Continuous Origin of Certain Unit Characters"; and *The Titanotheres of Ancient Wyoming, Dakota and Nebraska.*

70. See *The Titanotheres of Ancient Wyoming,* 2: 844.

71. See especially Osborn, *The Age of Mammals,* p. 29.

72. See Osborn, "The Four Inseparable Factors." The term "tetrakinesis" was first used in his *The Origin and Evolution of Life.*

73. *The Origin and Evolution of Life,* p. 150.

74. *The Titanotheres of Ancient Wyoming,* 2: 849. Note that Osborn used the term "phylum" in a much more restricted sense than the modern one.

75. The term "aristogenesis" literally means "the production of the best." It was first used in "The Nine Principles of Evolution." Osborn drew a clear connection with the teleological aspects of Aristotle's philosophy in "The Four Inseparable Factors of Evolution," p. 277.

76. See the letter from Morgan to Osborn, 26 December 1917, quoted in Garland

E. Allen, "T. H. Morgan and the Emergence of a New American Biology," pp. 179–80.

77. Kellogg, "Is there Determinate Variation?"

78. V. Jollos, "Studien zum Evolutionsproblem," supported by R. Goldschmidt, "Some Aspects of Evolution." For counterevidence see H. H. Plough and P. T. Ives, "Heat Induced Mutations in *Drosophila.*"

79. Shull, *Evolution*, pp. 123–33.

80. Goldschmidt, *The Material Basis of Evolution*, p. 322. On Goldschmidt's views, see Garland E. Allen, "Opposition to the Mendelian-Chromosome Theory."

81. On Simpson's work, see Stephen Gould, "G. G. Simpson, Paleontology, and the Modern Synthesis."

82. See J. B. S. Haldane, *The Causes of Evolution*, pp. 23–28; and on the dangers of intraspecific competition, pp. 125–28. For early efforts to reinterpret Hyatt's theory of racial senility, see Edward W. Berry, "Cephalopod Adaptations," and Carl Owen Dunbar, "Phases of Cephalopod Adaptation."

83. Lankester, "Are Specific Characters Useful?" (another response to Wallace's paper). See also Lankester's 1906 "President's Address," p. 31.

84. T. H. Morgan, *A Critique of the Theory of Evolution*, pp. 71–72.

85. A. H. Sturtevant, "An Interpretation of Orthogenesis."

86. Julian Huxley, "Constant Differential Growth-Ratios" and *Problems of Relative Growth*, pp. 214–21.

87. Francis A. Bather, "Fossils and Life," pp. 85–86.

88. See, for instance, Osborn, *Man Rises to Parnassus.*

89. Hyatt, "The Influence of Woman in the Evolution of the Human Race."

90. George W. Crile, "Orthogenesis and the Power and Infirmities of Man."

91. Bather, "Fossils and Life," p. 86.

CHAPTER 8

1. Huxley to Darwin, 23 November 1859, quoted in Leonard Huxley, *Life and Letters of Thomas Henry Huxley*, 1: 254. See also Huxley's review, "The Origin of Species," *Darwiniana*, p. 77.

2. Huxley, "The Origin of Species," *Darwiniana*, p. 38.

3. Huxley to Bateson, 20 February 1894, *Life and Letters of Huxley*, 3: 320. Huxley admitted that he had not had time to read *Materials* properly.

4. Francis Galton, *Hereditary Genius*, p. 369; 1892 ed., pp. 354–55. On Galton's work, see Pearson's biography and the articles by J. S. Wilkie, R. G. Swinburne, Ruth S. Cowan, and Robert De Marrais.

5. Galton, *Natural Inheritance*, p. 27.

6. On Galton's role as the founder of eugenics see, for instance, Mark Haller, *Eugenics;* G. R. Searle, *Eugenics and Politics in Britain*; and Carl Jay Bajema, *Eugenics, Then and Now.*

7. Mivart, *Genesis of Species*, p. 261.

8. See Poulton's 1898 paper, "Natural Selection and the Cause of Mimetic Resemblances," reprinted in Poulton, *Essays on Evolution, 1889-1907*, pp. 220-70. For a specific attack on Bateson's lack of field experience, see *Essays on Evolution*, preface.

9. Bateson, "The Ancestry of the Chordata," reprinted from *Quarterly Journal of Microscopic Science* (1886) in Bateson, *Scientific Papers*, 1: 1-31. On Bateson's career, see Beatrice Bateson, *William Bateson: Naturalist.*

10. See the letter by Gregory Bateson quoted in Koestler, *Case of the Midwife Toad*, p. 42. Koestler explains Bateson's later hostility to Lamarckism as a consequence of this early frustration.

11. Bateson, "On Some Variations of *Cardium edule* Apparently Correlated to the Conditions of Life," reprinted from *Philosophical Transactions of the Royal Society* (1889) in *Scientific Papers*, 1: 33-70; see p. 34.

12. See the letter quoted in Beatrice Bateson, *William Bateson: Naturalist*, p. 14.

13. Provine, *Origins of Theoretical Population Genetics*, pp. 42-43.

14. Bateson, "On the Variations in Floral Symmetry of Certain Plants having Irregular Corollas," reprinted from *Journal of the Linnaean Society* (1891) in *Scientific Papers*, 1: 126-61; see p. 129.

15. Bateson, *Materials for the Study of Variation*, p. 11.

16. Ibid., p. 15.

17. Ibid., pp. 15-16.

18. Ibid., pp. 7 and 17.

19. Ibid., p. 76.

20. On the debate between Mendelism and biometry see, for instance, Provine, *Origins of Theoretical Population Genetics;* P. Froggatt and N. C. Nevin, "The 'Law of Ancestral Heredity' and the Mendelian-Ancestrian Controversy"; and B. J. Norton, "The Biometric Defense of Darwinism."

21. Bateson, *Problems of Genetics* (1979 reprint ed.), p. xxi. In a 1921 address to the A.A.A.S. in Toronto, Bateson again expressed this pessimism and was taken to task by H. F. Osborn for overstating the case so that many thought he had rejected the basic idea of evolution; see Bateson, "Evolutionary Faith and Modern Doubts," and Osborn, "William Bateson on Darwinism."

22. This anticipation of Goldschmidt's approach is suggested by G. Evelyn Hutchinson and Stan Rachootin in their introduction to Bateson's *Problems of Genetics*, p. xix.

23. Ibid., p. 35.

24. On Bateson's reasons for rejecting the chromosome theory, see William Coleman, "Bateson and Chromosomes: Conservative Thought in Science"; and on the link with the physicists, see Garland Allen, *Life Sciences in the Twentieth Century*, p. 57.

25. Bateson, *Problems of Genetics*, p. 36. The vibration theory of heredity was already sketched out in an 1891 letter to his sister, in which Bateson claimed an eight-petaled flower was related to a four-petaled form as a note is to its octave. See B. Bateson, *William Bateson: Naturalist*, p. 43.

26. *Problems of Genetics*, p. 60.

27. Ibid., p. 50.

28. C. H. Waddington suggests that Bateson's hostility arose from the fact that he could not accept the possibility of a simple misinterpretation of the results. The observed effect could have been explained as the result of unconscious selection in the experimental population—but since Bateson was incapable of thinking in populational terms, he could not allow Kammerer this excuse. Lamarckism was the only possible explanation; therefore, the results had to be fraudulent. See Waddington, *The Evolution of an Evolutionist*, pp. 177–78.

29. *Problems of Genetics*, pp. 219–27.

30. Ibid., p. 191.

31. Ibid., p. 90.

32. Bateson, "President's Address," pp. 12, 16, 20. See also *Problems of Genetics*, pp. 94–96.

33. "President's Address," pp. 17–18.

34. Bateson, "Heredity and Variation in Modern Light," p. 109.

35. See, for instance, R. Ruggles Gates, *The Mutation Factor in Evolution*, pp. 313–14.

36. Bateson, *Problems of Genetics*, pp. 102–3, although he had made the same point on earlier occasions.

37. J. P. Lotsy, *Evolution by Means of Hybridization*, p. 53.

38. Ibid., pp. 132–36.

39. Ibid., pp. 118–19.

40. Ibid., p. 135.

41. Bateson, "President's Address," p. 15.

42. Loren Eiseley, *Darwin's Century*, pp. 250–51.

43. See Lindley Darden, "Reasoning in Scientific Change: Charles Darwin, Hugo De Vries, and the Discovery of Segregation," and Malcolm Kottler, "Hugo De Vries and the Rediscovery of Mendel's Laws."

44. The translation of *The Mutation Theory* did not appear until 1910. On the reception of the theory in the English-speaking world, see Garland Allen, "Hugo De Vries and the Reception of the Mutation Theory"; see also Bowler, "Hugo De Vries and Thomas Hunt Morgan: The Mutation Theory and the Spririt of Darwinism."

45. De Vries, *Species and Varieties*, p. ix.

46. De Vries, *The Mutation Theory*, 1: 35–39; see also *Species and Varieties*, p. 6.

47. For the most concise statement of the laws governing mutation, see *Species and Varieties*, pp. 556–75.

48. *The Mutation Theory*, 1: 85–92; also *Species and Varieties*, pp. 770–97.

49. *The Mutation Theory*, 1: 65.

50. Ibid., p. 6; also *Species and Varieties*, p. 430.

51. *The Mutation Theory*, 1: 205–7; also *Species and Varieties*, pp. 686–714.

52. *The Mutation Theory*, 1: 203.

53. Ibid., pp. 57–59.

54. Ibid., p. 198; also *Species and Varieties*, p. 571.

55. *The Mutation Theory*, 1: 204.

56. Ibid., p. 66, criticizing Scott, "On Variations and Mutations."

57. Thomas L. Casey, "The Mutation Theory."

58. De Vries, *The Mutation Theory*, 1: 157, quoting Otto Ammon, *Die Gesellschaftsordnung und ihre näturlichen Grundlagen* (1896).

59. Ibid., pp. 155-56.

60. On Morgan's views, see Allen, "Thomas Hunt Morgan and the Problem of Natural Selection," "T. H. Morgan and the Emergence of a New American Biology," and *Thomas Hunt Morgan: The Man and His Science*. See also Bowler, "Hugo De Vries and Thomas Hunt Morgan."

61. Morgan, *Evolution and Adaptation*, p. 23.

62. Ibid.

63. Ibid., p. 368.

64. Ibid., pp. 348-49.

65. Ibid., pp. 119-21.

66. Ibid., p. 351.

67. Ibid., pp. 325-39.

68. Ibid., p. 116.

69. For example, Thomas Dwight, "Mutations."

70. Maynard Metcalf, "Determinate Mutation."

71. Gates, *The Mutation Factor in Evolution*, p. 303.

72. Ibid., pp. 6 and 322.

73. See Bradley Davis, "Genetical Studies on *Oenothera*."

74. Clelland's work began in 1923 with "Chromosome Arrangements during Meiosis in Certain *Oenotheras*." Morgan's essentially modern interpretation of genetics and mutation was outlined in major works such as *The Theory of the Gene*. See Allen's biography for details.

75. Morgan, *Critique of the Theory of Evolution*, p. 190.

76. C. Hart Merriam, "Is Mutation a Factor in the Evolution of the Higher Vertebrates?"

77. David Starr Jordan, "The Origin of Species through Isolation," pp. 549-50.

78. J. A. Allen, "Mutations and the Geographic Distribution of Nearly Related Species."

79. Zittel's address was published under the title "Ontogenie, Phylogenie und Systematik" in the conference proceedings; see his "Palaeontology and the Biogenetic Law." The most accessible accounts of his views are in Depéret's *Transformations of the Animal World*, chap. 12, and Russell's *Form and Function*, pp. 357-58.

80. O. H. Schindewolf, *Paläontologie Entwickelungslehre und Genetik*, p. 59. On the work of Schindewolf and the other paleontologists supporting mutation, see

B. Rensch, "Historical Development of the Present Synthetic Neo-Darwinism in Germany," pp. 289-90.

81. Edwin G. Conklin, "The Mutation Theory from the Standpoint of Cytology," p. 529.

82. Richard Goldschmidt, *The Material Basis of Evolution*, p. 322. On the "hopeful monster," see pp. 390-93. Goldschmidt and his co-worker V. Jollos claimed to have evidence for parallel mutations induced in *Drosophila* by heat; see Goldschmidt, "Some Aspects of Evolution." For a critique of these experiments from the orthodox viewpoint, see H. H. Plough and P. T. Ives, "Heat Induced Mutations in *Drosophila*." For a survey of Goldschmidt's alternative interpretation of genetics, see Garland Allen, "Opposition to the Mendelian-Chromosome Theory." The views of Schindewolf and Goldschmidt are now quoted seriously by, for instance, Søren Løvtrup, a leading advocate of the "epigenetic" approach to evolution; see *The Phylogeny of the Vertebrata*.

83. H. B. Guppy, *Observations of a Naturalist in the Pacific*, 2: 522-23.

84. Guppy, *Plants, Seeds and Currents in the West Indies*, pp. 313-22.

85. Guppy, "Plant Distribution from the Standpoint of an Idealist," p. 445.

86. J. C. Willis, *Age and Area*, chap. 3. Views somewhat similar to those of Willis on the origin of new types were advocated under the name "hologenesis" by the Italian zoologist Daniele Rosa in 1918. Despite a translation of his work into French in 1931, Rosa's views were either ignored or dismissed by the supporters of the modern synthesis. For a brief account of this episode, see Cesare Baroni-Urbani, "Hologenesis, Phylogenetic Systematics, and Evolution."

87. De Vries, in Willis, *Age and Area*, chap. 21, p. 227.

88. Willis, *The Course of Evolution*, p. 178.

89. Ibid., p. 191.

90. Ibid., p. 166.

91. Ibid., p. 72.

92. R. C. Punnett, *Mimicry in Butterflies*, p. 62. For more details of Punnett's views, see Provine, *Origins of Theoretical Population Genetics*, pp. 137-39.

93. R. A. Fisher, "Some Objections to Mimicry Theory." On Fisher's work, see the biography by Joan Fisher Box.

94. T. H. Morgan, *A Critique of the Theory of Evolution*, p. 67.

95. Ibid., pp. 71-72.

96. Julian Huxley, *Evolution: The Modern Synthesis*, p. 417.

97. See Garland Allen, "Thomas Hunt Morgan and the Problem of Natural Selection," pp. 136-37; and, in general, Allen, *Thomas Hunt Morgan*, and Bowler, "Hugo De Vries and Thomas Hunt Morgan."

98. Morgan, *The Scientific Basis of Evolution*, pp. 131 and 150.

99. Morgan, *Critique of the Theory of Evolution*, pp. 87-88.

100. Robson and Richards, *The Variations of Animals in Nature*, chap. 7, especially p. 316.

101. Ibid., chap. 10.

102. Charles Elton, *Animal Ecology*, chap. 12. Elton was a student of Julian Huxley. He acknowledged that selection must be effective in the long run.

103. On the development of the modern synthesis see, for instance, Provine, *Origins of Theoretical Population Genetics,* and Mayr and Provine, eds., *The Evolutionary Synthesis.*

104. For a survey of the punctuationist position, see Stephen Gould and Niles Eldredge, "Punctuated Equilibria: The Tempo and Mode of Evolution Reconsidered."

105. For discussion of these new ideas, see Stephen J. Gould, "Is a New and General Theory of Evolution Emerging?" and "Punctuated Equilibrium—a Different Way of Seeing"; Stephen J. Gould and R. C. Lewontin, "The Spandrels of San Marco and the Panglossian Paradigm"; Stan Rachootin and Keith Thomson, "Epigenetics, Paleontology, and Evolution"; and Søren Løvtrup, *The Phylogeny of the Vertebrata.*

106. For an evolutionist's assessment of the cladists' attack, see David Hull, "The Limits of Cladism." For a more general response to the modern attacks on Darwinism, see Michael Ruse, *Darwinism Defended.*

BIBLIOGRAPHY

This is by no means a complete record of all the pro- and anti-Darwinian literature published in the late nineteenth and early twentieth centuries. It may be supplemented by the bibliographies given in a number of the works listed below, especially R. Ruggles Gates's *Mutation Factor in Evolution,* Glenn L. Jepsen's "Selection, 'Orthogenesis,' and the Fossil Record," Paul Kammerer's *Inheritance of Acquired Characteristics,* and J. Arthur Thomson's *Heredity.*

I thought it useful to separate the recent secondary literature into a section of its own. Note, however, that some older works that would normally be listed as secondary literature are actually primary sources for this debate and are listed as such below, for example, Eric Nordenskiöld's *History of Biology.* Some contributors to the modern synthesis have written both scientific and historical material in more recent years; here I have made an arbitrary decision to list anything written by a working scientist after 1945 as a secondary source. "Lives and Letters" and obituaries are listed as primary sources.

PRIMARY SOURCES

Agassiz, Alexander. "Paleontological and Embryological Development." *Am. J. Sci.* 120 (1880): 294-302, 375-89.

Agassiz, Elizabeth Cary. *Louis Agassiz: His Life and Correspondence.* 2 vols. London: Macmillan, 1885.

Agassiz, G. R. *Letters and Recollections of Alexander Agassiz: With a Sketch of his Life and Work.* Boston: Houghton Mifflin, 1913.

Agassiz, Louis. "Evolution and Permanence of Type." *Atlantic Monthly* 33 (1874): 92-101.

———. *Essay on Classification,* ed. Edward Lurie. Cambridge: Harvard University Press, 1962. Reprint of vol. 1 of *Contributions to the Natural History of the United States,* 1857.

Alexander, Samuel. *Space, Time and Deity.* 2 vols. London: Macmillan, 1920.

Allen, Joel A. "Geographical Variation in North American Birds." *Proc. Boston Soc. Nat. Hist.* 15 (1872): 212-19.

———. "Remarks on the Geographical Variation of Mammals and Birds." *Proc. Boston Soc. Nat. Hist.* 15 (1872): 156-59.

———. "The Influence of Physical Conditions in the Genesis of Species." *Radical Review* 1 (1877-78): 108-37.

———. "Mutations and the Geographic Distribution of Nearly Related Species in Plants and Animals." *Am. Naturalist* 41 (1907): 653-55.

Argyll, Dowager Duchess of. *George Douglas Campbell, Eighth Duke of Argyll, K.G., K.T. (1823-1900): Autobiography and Memoirs.* 2 vols. London: Murray, 1906.

Argyll, George Douglas Campbell, 8th Duke of. *The Reign of Law.* 5th ed. London: Alexander Strahan, 1868.

——. "On Variety as an Aim in Nature." *Contemporary Review* 17 (1871): 153-60.

——. *The Unity of Nature.* London: Alexander Strahan, 1884.

——. "Organic Evolution." *Nature* 34 (1886): 335-36.

——. "Prophetic Germs." *Nature* 38 (1888): 564.

——."Acquired Characters and Congenital Variations." *Nature* 41 (1888-89): 173-74.

——. *Organic Evolution Cross-Examined.* London: Murray, 1898.

Arrhenius, Svante. *Worlds in the Making: The Evolution of the Universe,* trans. H. Borns, New York: Harper, 1908.

Baer, Karl Ernst von. *Über Entwickelungsgeschichte der Thiere: Beobachtung und Reflexion. Erster Theil.* Königsberg, 1828. Reprint. Brussels: Culture et Civilisation, 1967.

Baldwin, James Mark. *Development and Evolution: Including Psychophysical Evolution, Evolution by Orthoplasy and the Theory of Genetic Modes.* New York: Macmillan, 1902.

Ball, W. Platt. *Are the Effects of Use and Disuse Inherited? An Examination of the View Held by Spencer and Darwin.* London: Macmillan, 1890.

Bates, Henry Walter. "Contributions to an Insect Fauna of the Amazon Valley. Lepidoptera: Heliconidae." *Trans. Linn. Soc. Lond.* 23 (1862): 495-515.

——. *The Naturalist on the River Amazons.* 2 vols. London: Murray, 1863.

Bateson, Beatrice. *William Bateson: Naturalist.* Cambridge: Cambridge University Press, 1928.

Bateson, William. *Materials for the Study of Variation: Treated with Especial Regard to Discontinuity in the Origin of Species.* London: Macmillan, 1894.

——. *Mendel's Principles of Heredity: A Defence.* Cambridge: Cambridge University Press, 1902.

——. *Mendel's Principles of Heredity.* 1909. Reprint. Cambridge: Cambridge University Press, 1913.

——. *Problems of Genetics.* 1913. Reprint, intro. G. Evelyn Hutchinson and Stan Rachootin. New Haven: Yale University Press, 1979.

——. "President's Address." *Report of the British Association for the Advancement of Science,* 1914 meeting, pp. 3-38.

——. "Heredity and Variation in Modern Light." In *Evolution in Modern Thought,* by Ernst Haeckel, J. Arthur Thomson, August Weismann, and others, 87-110. New York: Boni & Liveright, 1917.

——. "Evolutionary Faith and Modern Doubts."*Science* 55 (1922): 55-61.

——. *The Scientific Papers of William Bateson,* ed. R. C. Punnett. 2 vols. Cambridge: Cambridge University Press, 1928.

Bather, Francis A. "Pseudo-Biology." *Natural Science* 5 (1894): 449-54.

——. "Fossils and Life." *Report of the British Association for the Advancement of Science,* 1920 meeting, pp. 61-86.

Beecher, Charles E. *Studies in Evolution.* New York: Scribners, 1901.

Bell, Alexander Graham. "On the Formation of a Deaf Variety of the Human Race." *Mem. Nat. Acad. Sci.* 2, pt. 4 (1884): 177-262.

Bennet, A. W. "The Theory of Natural Selection from a Mathematical Point of View." *Nature* 3 (1870-71): 30-33.

Berg, Leo S. *Nomogenesis: Or Evolution Determined by Law,* trans. J. N. Rostov-tsov; intro. D'Arcy Wentworth Thompson. 1926. Reprint, intro. Theodosius Dobzhansky. Cambridge: MIT Press, 1969.

Bergson, Henri. *Creative Evolution,* trans. Arthur Mitchell. New York: Henry Holt, 1911.

Berry, Edward W. "Cephalopod Adaptations—The Record and its Interpretation." *Quart. Rev. Biol.* 3 (1928): 92–108.

Bonnier, Gaston. "Recherches expérimentales sur l'adaptation des plantes au climat alpin." *Ann. des sci. nat., botanique,* 7ième série, 20 (1895): 217–360.

———. "Expériences sur la production des chaactères alpins des plantes, par l'alter-nance des températures extrêmes." *Comptes rendus à l'Academie des Sciences* 127 (1898): 307–12.

———. "Charactères anatomiques et physiologiques des plantes rendues artificielle-ment alpines par l'alternance des températures extrêmes." *Comptes rendus à l'Academie des Sciences* 128 (1899): 1143–46.

———. "Cultures expérimentales sur l'adaptation des plantes au climat méditerrane-an," *Comptes rendus à l'Academie des Sciences* 139 (1899): 1207–13.

Brooks, William Keith. *The Law of Heredity: A Study of the Cause of Variation and the Origin of Living Organisms.* Baltimore: John Murphy, 1883.

———. "Biographical Memoir of Alpheus Hyatt (1838–1902)." *Biog. Mem. Nat. Acad. Sci.* 6 (1909): 311–25.

———. *The Foundations of Zoology.* 2d ed., rev. New York: Columbia University Press, 1915.

Brown-Séquard, Charles Édouard. "Hereditary Transmission of an Epileptiform Affection, Accidentally Produced." *Proc. Roy. Soc. Lond.* 10 (1860): 297–98.

———. "On the Hereditary Transmission of Effects of Certain Injuries to the Nervous System." *Lancet* I (1875): 7–8.

———. "Faits nouveaux établissant l'extrême frequence de la transmission par l'hérédité, d'états organiques morbides produit accidentellement chez des as-cendents." *Comptes rendus à l'Academie des Sciences* 94 (1882): 697–700.

———. "Hérédité d'une affection due a une cause accidentelle. Faits et arguments contre les explications et les critiques de Weismann." *Arch. physiol.* 24 (1892): 686–88.

———. "Transmission héréditaire des charactères acquis." *Arch. physiol.* 25 (1893): 209–10.

Butler, Samuel. *Evolution, Old and New: Or the Theories of Buffon, Dr. Erasmus Darwin, and Lamarck, as Compared with that of Mr. Charles Darwin.* London: Hardwick and Bogue, 1879.

———. *Essays on Life, Art and Science,* ed. R. A. Streatfield. London, 1908. Reprint. Port Washington, N.Y.: Kennikat Press, 1970.

———. *Life and Habit.* New ed. London: A. C. Fifield, 1916.

———. *Unconscious Memory.* 3d ed. London: A. C. Fifield, 1920.

———. *Luck, or Cunning, as the Main Means of Organic Modification?* 2d ed. Lon-don: A. C. Fifield, 1920.

———. *The Note-Books of Samuel Butler,* ed. Henry Festing Jones. London: Jona-than Cape, 1930.

Carpenter, William Benjamin. "On the Hereditary Transmission of Acquired Psych-ical Habits." *Contemporary Review* 21 (1873): 295–314, 779–95, 867–85.

———. *Mental Evolution in Animals.* London: Kegan, Paul, Trench and Co., 1883.

Carpenter, William Benjamin. *Nature and Man: Essays Scientific and Philosophical. With an Introductory Memoir by J. Erstlin Carpenter.* New York: Appleton, 1889.

Casey, Thomas L. "The Mutation Theory." *Science,* n.s. 21 (1905), pp. 307-9.

Castle, W. E., "The Mutation Theory of Organic Evolution, from the Standpoint of Animal Breeding." *Science,* n.s. 21 (1905): 521-25.

Caullery, Maurice. *Le problème de l'évolution.* Paris: Payot, 1931.

Chambers, Robert. *Vestiges of the Natural History of Creation.* London: 1844. Reprint, intro. Sir Gavin De Beer. Leicester: Leicester University Press, 1969.

——. *Explanations: A Sequel to the Vestiges of the Natural History of Creation.* 2d ed. London: John Churchill, 1846.

——. *Vestiges of the Natural History of Creation.* 11th ed. London: John Churchill, 1860. 12th ed., with intro. by Alexander Ireland. Edinburgh: W. & R. Chambers, 1884.

Clelland, R. E. "Chromosome Arrangements during Meiosis in Certain *Oenotheras.*" *Am. Naturalist* 57 (1923): 562-66.

Cockerell, T. D. A. "Biographical Memoir of Alpheus Spring Packard." *Biog. Mem. Nat. Acad. Sci.* 9 (1920): 179-236.

Conklin, Edwin G. "The Mutation Theory from the Standpoint of Cytology." *Science,* n.s. 21 (1905): 525-29.

Conn, H. W. *The Method of Evolution. A Review of the Present Attitude of Science Toward the Question of the Laws and Forces which have Brought about the Origin of Species.* New York: Putnam, 1900.

Cope, Edward Drinker. "On the Origin of Genera." *Proc. Acad. Nat. Sci., Philadelphia* 20 (1868): 242-300.

——. "The Laws of Organic Development." *Am. Naturalist* 5 (1871): 593-605.

——. "The Method of Creation of Organic Forms." *Proc. Am. Phil. Soc.* 12 (1873): 229-65.

——. *The Origin of the Fittest: Essays in Evolution.* New York: Macmillan, 1887a. Reprinted with Cope, *Primary Factors of Organic Evolution.* New York: Arno Press, 1974.

——. *Theology of Evolution: A Lecture.* Philadelphia: Arnold & Co., 1887b.

——. "The Mechanical Origins of the Hard Parts of the Mammalia." *Am. Naturalist* 23 (1889): 71-73.

——. "The Energy of Evolution." *Am. Naturalist* 28 (1894): 205-19.

——. *The Primary Factors of Organic Evolution.* Chicago: Open Court, 1896.

Crampton, Henry Edward. *The Doctrine of Evolution: Its Basis and Scope.* New York: Columbia University Press, 1912.

Crile, George W. "Orthogenesis and the Power and Infirmities of Man." *Proc. Am. Phil. Soc.* 72 (1933): 245-54.

Cuénot, Lucien. *La genèse des espèces animales.* Paris: Alcan, 1911.

——. *L'évolution biologique: Les faits, les incertitudes.* Paris: Masson et cie., 1951.

Cunningham, Joseph T. "The New Darwinism." *Westminster Review* 136 (1891): 14-28.

——. "Blind Animals in Caves." *Nature* 47 (1892-93): 439.

——. "The Problem of Variation." *Natural Science* 3 (1893): 282-87.

——. "The Origin and Evolution of Flatfishes." *Natural Science* 6 (1895): 169-77, 233-39.

——. "The Utility of Specific Characters." *Nature* 54 (1896): 295.

——. *Sexual Dimorphism in the Animal Kingdom: A Theory of the Origin of Secondary Sexual Characters.* London: A. & C. Black, 1900.

——. *Hormones and Heredity: A Discussion of the Evolution of Adaptations and the Evolution of Species.* London: Constable, 1921.

——. *Modern Biology: A Review of the Principal Phenomena of Animal Life in Relation to Modern Concepts and Theories.* London: Kegan Paul, Trench, Trubner, 1928.

——, and MacMunn, C. A. "On the Coloration of the Skins of Fishes, Especially the Pleuronectidae." *Phil. Trans. Roy. Soc. Lond.* (B), 184 (1893): 765–812.

Darwin, Charles Robert. *On the Origin of Species by Means of Natural Selection: Or the Preservation of Favoured Races in the Struggle for Life.* London, 1859. Reprint, intro. Ernst Mayr. Cambridge: Harvard University press, 1964. 6th ed. London: Murray, 1872.

——. *The Variation of Animals and Plants under Domestication.* 2 vols. 2d ed. London: Murray, 1882.

——. *The Descent of Man: And Selection in Relation to Sex.* 2d ed., rev. London: Murray, 1885.

——. *The Collected Papers of Charles Darwin,* ed. Paul H. Barrett. 2 vols. Chicago: University of Chicago Press, 1977.

Darwin, Francis, ed. *The Life and Letters of Charles Darwin.* 3 vols. London: Murray, 1887.

——. *More Letters of Charles Darwin: A Record of his Life and Work in a Series of Unpublished Letters.* 2 vols. New York: Appleton, 1903.

——. "President's Address." *Report of the British Association for the Advancement of Science,* 1908 meeting, pp. 3–27.

Davenport, Charles Benedict. "The Mutation Theory in Animal Evolution." *Science,* n.s. 24 (1907): 556–58.

——. *Heredity in Relation to Eugenics.* 1911. Reprint, intro. Charles E. Rosenberg. New York: Arno Press, 1972.

Davis, Bradley. "Genetical Studies on *Oenothera.* I. Notes on the Behavior of Certain Hybrids of *Oenothera* in the First Generation." *Am. Naturalist* 44 (1910): 108–15.

Dawson, Sir John William. *Life's Dawn on Earth: Being the History of the Oldest Known Fossil Remains and their Relations to Geological Time and the Development of the Animal Kingdom.* London: Copp, Clark, 1875.

——. *Modern Ideas of Evolution.* 1890. Reprint, ed. William R. Shea and John F. Cornell. New York: Prodist, 1977.

Dean, Bashford. "Evolution in a Determinate Line, as Illustrated by the Egg-cases of Chimaeroid Fishes." *Biological Bulletin* 7 (1904): 105–12.

De Beer, Gavin R. *Embryology and Evolution.* Oxford: Oxford University Press, 1930.

——. *Embryos and Ancestors.* Oxford: Oxford University Press, 1940.

Delage, Yves. *L'hérédité et les grands problèmes de la biologie générale.* 2d ed. Paris: Schleicher frères, 1903.

——, and Goldsmith, Marie. *The Theories of Evolution,* trans. André Tridon. New York: Huebsch, 1912.

Delcourt, A., and Guyenot, E. "Genetique et milieu: Necessité de la détermination des conditions. Sa possibilité chez les Drosophiles." *Bulletin scientifique de la France et de la Belgique* 45 (1911): 249–322.

Demoor, Jean; Massart, Jean; and Vandervelde, Émile. *Evolution by Atrophy in Biology and Sociology,* trans. Mrs. Chalmers Mitchell. New York: Appleton, 1899.

Dendy, Arthur. "Momentum in Evolution." *Report of the British Association for the Advancement of Science*, 1911 meeting, pp. 277-80.

———. "Progressive Evolution and the Origin of Species." *Report of the British Association for the Advancement of Science*, 1914 meeting, pp. 383-97.

Dennert, Eberhart. *At the Deathbed of Darwinism*, trans. E. V. O'Harra and John H. Peschges. Burlington, Iowa: German Literary Board, 1904.

Depéret, Charles. *Les transformations du monde animal*. Paris: E. Flammarion, 1907.

———. *The Transformations of the Animal World*. London: Kegan Paul, Trench, Trubner, 1909.

Detlefsen, J. A. "Are the Effects of Long-Continued Rotation in Rats Inherited?" *Proc. Am. Phil. Soc.*, 42 (1923): 292-300.

De Vries, Hugo. *Species and Varieties: Their Origin by Mutation*, ed. D. T. Mac-Dougal. Rev. ed. Chicago: Open Court, 1906.

———. *Intracellular Pangenesis: Including a Paper on Fertilization and Hybridization*, trans. C. Stuart Gager. Chicago: Open Court, 1910a.

———. *The Mutation Theory: Experiments and Observations on the Origin of Species in the Vegetable Kingdom*, trans. J. B. Farmer and A. D. Darbyshire. 2 vols. London: Kegan Paul, Trench, Trubner, 1910b.

Dewey, John. *The Influence of Darwin on Philosophy and Other Essays in Contemporary Thought*. 1910. Reprint. Bloomington: Indiana University Press, 1965.

Dobzhansky, Theodosius. *Genetics and the Origin of Species*. New York: Columbia University Press, 1937.

Döderlein, Ludwig. "Phylogenetische Betrachtungen." *Biologisches Centralblatt* 7 (1888): 394-402.

Driesch, Hans. *The Science and Philosophy of the Organism. The Gifford Lectures Delivered before the University of Aberdeen in the Years 1907 and 1908*. 2 vols. London: A. & C. Black, 1908.

Drummond, Henry. *The Ascent of Man*. 13th ed. New York: James Pott, 1904.

Dunbar, Carl Owen. "Phases of Cephalopod Adaptation." In *Organic Adaptation to the Environment*, ed. M. R. Thorpe, 188-223. New Haven: Yale University Press, 1924.

Duncan, David. *The Life and Letters of Herbert Spencer*. Reissued. London: Methuen, 1911.

Dwight, Thomas. "Mutations." *Science*, n.s. 21 (1905): 529-32.

Eimer, Gustav Heinrich Theodor. *Zoologische Studien auf Capri. Vol. 2. Lacerta muralis caerula. Ein Beitrag zur Darwin'schen Lehre*. Leipzig: Engelmann, 1874.

———. *Die Entstehung der Arten auf Grund von Vererben erworbener Eigenschaften nach den Gesetzen organischen Wachsens*. Jena: G. Fischer, 1888.

———. *Organic Evolution as the Result of the Inheritance of Acquired Characters According to the Laws of Organic Growth*, trans. J. T. Cunningham. London: Macmillan, 1890.

———. *Die Entstehung der Arten . . . zweiter Theil. Orthogenesis der Schmetterlinge. Ein Beweis bestimmt gerichteter Entwickelung und Ohnmacht der natürlichen Zuchtwahl bei der Artbildung*. Leipzig: Engelmann, 1897.

———. *On Orthogenesis and the Impotence of Natural Selection in Species-Formation*, trans. Thomas J. McCormack. Chicago: Open Court, 1898.

Elton, Charles. *Animal Ecology*, Intro. Julian Huxley. London: Sidgwick and Jackson, 1927.

Emmet, Dorothy. *Whitehead's Philosophy of Organism.* London, 1932. New ed. London: Macmillan, 1966.

Fisher, Ronald Aylmer. "The Correlation between Relatives on the Supposition of Mendelian Inheritance." *Trans. Roy. Soc. Edinburgh* 52 (1918): 399–433.

———. "On Some Objections to Mimicry Theory: Statistical and Genetic." *Trans. Ent. Soc. Lond.* 75 (1927): 269–78.

———. *The Genetical Theory of Natural Selection.* Oxford: Clarendon Press, 1930.

Fry, Edward. "What are Acquired Characters?" *Nature* 51 (1894-95): 8–11.

Galton, Sir Francis. *Hereditary Genius: An Inquiry into its Laws and Consequences.* London, Macmillan, 1869. 1892 rev. ed. reprint, intro. C. D. Darlington. Cleveland: Meridian Books, 1962.

———. *Natural Inheritance.* London: Macmillan, 1889.

———. "Discontinuity in Evolution." *Mind,* n.s. 3 (1894): 362–72.

Gates, R. Ruggles. *The Mutation Factor in Evolution: With Particular Reference to Oenothera.* London: Macmillan, 1915.

Geddes, Patrick and Thomson, J. Arthur. *The Evolution of Sex.* London: Walter Scott, 1889.

———. *Evolution.* 1911. Reprint. London: Williams and Norgate, 1924.

Giard, Alfred. "L'évolution des êtres organisés: Leçon d'ouverture." *Bulletin scientifique* 20 (1889a): 1–26.

———. "Les facteurs de l'évolution." *Revue scientifique* 44 (1889b): 641–48.

———. "Le principe de Lamarck et l'hérédité des modifications somatiques." *Revue scientifique* 46 (1890): 705–13.

———. *Controverses transformistes.* Paris: Naud, 1904.

Gill, Theodor. "Edward Drinker Cope, Naturalist—A Chapter in the History of Science." *Am. Naturalist* 31 (1897): 831–63.

Goldschmidt, Richard. "Some Aspects of Evolution." *Science* 78 (1933): 539–47.

———. *The Material Basis of Evolution.* 1940. Reprint. Paterson, N. J.: Pageant Books, 1960.

Goodrich, E. T. *Living Organisms: An Account of their Origin and Evolution.* Oxford: Oxford University Press, 1924.

Gray, Asa. *Darwiniana: Essays and Reviews Pertaining to Darwinism.* New York, 1876. Reprint, ed. A. Hunter Dupree. Cambridge: Harvard University Press, 1963.

Gregory, William K. "Biographical Memoir of Henry Fairfield Osborn." *Biog. Mem. Nat. Acad. Sci.* 19 (1938): 53–119.

Gulick, Addison. *Evolutionist and Missionary: John Thomas Gulick.* Chicago: University of Chicago Press, 1932.

Gulick, John Thomas. "Divergent Evolution through Cumulative Segregation." *J. Linn. Soc. (Zool.)* 20 (1888): 189–274.

Guppy, H. B. *Observations of a Naturalist in the Pacific between 1896-1899.* 2 vols. London: Macmillan, 1906.

———. *Plants, Seeds and Currents in the West Indies.* London: Williams and Norgate, 1917.

———. "Plant Distribution from the Standpoint of an Idealist." *J. Linn. Soc. (Bot.)* 44 (1917-20): 439–72.

Haacke, Wilhelm. *Gestaltung und Vererbung: Ein Entwickelungsmechanik der Organismen.* Leipzig: Weigel, 1893.

Haeckel, Ernst. *Generelle Morphologie der Organismen: Allgemeine Grundzüge der organischen Formenwissenschaft, mechanisch begründet durch die von Ch. Darwin reformierte Descenenztheorie.* Berlin: Georg Reimer, 1866.

Haeckel, Ernst. *Natürliche Schöpfungsgeschichte, gemeinverständliche wissenschaftliche Vorträge über die Entwickelungslehre in Allgemeinen und die jenige von Darwin, Goethe und Lamarck im Besonderen.* Berlin: Reimer, 1873.

———. *Anthropogenie; oder Entwickelungsgeschichte des Menschens . . .* Leipzig: Engelmann, 1874.

———. *The History of Creation: Or the Development of the Earth and its Inhabitants by the Action of Natural Causes. A Popular Exposition of the Doctrine of Evolution in General and of that of Darwin, Goethe and Lamarck in Particular,* trans. E. Ray Lankester. 2 vols. New York: Appleton, 1876a.

———. *Die Perigenesis der Plastidule oder die Wellenzeugung der Lebenstheilchen.* Berlin: Reimer, 1876b.

———. *The Evolution of Man: A Popular Exposition of the Principal Points of Human Ontogeny and Phylogeny.* New York: Appleton, 1879. 5th ed., London: Watts, 1907.

———. *The Last Link: Our Present Knowledge of the Descent of Man.* London: A. & C. Black, 1898.

———. *The Riddle of the Universe at the Close of the Nineteenth Century.* London: Watts, 1900.

Haeckel, Ernst; Thomson, J. Arthur; Weismann, August; and others. *Evolution in Modern Thought.* New York: Boni & Liveright, 1917.

Haldane, J.B.S. *Possible Worlds and Other Essays.* London: Chatto & Windus, 1927.

———. *The Causes of Evolution.* London, 1932a. Reprint. Ithaca, N. Y.: Cornell University Press, 1966.

———. "The Hereditary Transmission of Acquired Characters." *Nature* 129 (1932b): 817–19, 856–58.

———. *Heredity and Politics.* London: Allen & Unwin, 1938. Reprint: New York: Norton, 1938.

Harrison, J. W. Heslop. "Experiments on the Egg-laying Instincts of the Saw-fly, *Pontania salicis,* Christ., and their Bearing on the Inheritance of Acquired Characters, with some Remarks on a New Principle of Evolution." *Proc. Roy. Soc. Lond.* (B), 101 (1927): 115–26.

———, and Garrett, F. C. "The Induction of Melanism in the Lepidoptera and its Subsequent Inheritance." *Proc. Roy. Soc. Lond.* (B), 99 (1925–26): 241–63.

Hartog, Marcus. "The Fundamental Principles of Heredity." *Natural Science* 11 (1897): 233–39, 305–16.

———. "Samuel Butler and Recent Mnemic Biological Theories." *Scientia* 15 (1914): 38–52.

Henderson, Lawrence J. *The Fitness of the Environment: An Inquiry into the Biological Significance of the Properties of Matter.* 1913. Reprint, intro. George Ward. Gloucester, Mass.: Peter Smith, 1970.

———. "Orthogenesis from the Standpoint of the Biochemist." *Am. Naturalist* 56 (1922): 97–104.

Henslow, George. *The Origin of Floral Structures through Insect and Other Agencies.* London: Kegan Paul, 1888.

———. "The Origin of Plant Structures by Self-Adaptation to the Environment, Exemplified by Desert or Xerophilous Plants." *J. Linn. Soc. (Bot.)* 30 (1894a): 218–63.

———. "The Origin of Species without the Aid of Natural Selection—a Reply." *Natural Science* 5 (1894b): 257–64.

———. *The Origin of Plant Structures by Self-Adaptation to the Environment.* London: Kegan Paul, 1895.

——. "Does Natural Selection Play Any Part in the Origin of Species Among Plants?" *Natural Science* 11 (1897): 166–80.

——. "Scientific Proofs vs. '*A Priori*' Assumptions." *Natural Science* 13 (1898): 103–8.

Hertwig, Oscar. *Die Zelle und die Gewebe: Grundzüge der allgemeinen Anatomie und Physiologie.* 2 vols. Jena: G. Fischer, 1893–98.

——.*Das Werden der Organismen: Eine Widerlegung von Darwin's Zufalls-Theorie.* Jena: G. Fischer, 1916.

Hogben, Lancelot T. "The Progressive Reduction of the Jugal in The Mammalia." *Proc. Zool. Soc. Lond.* (1919): 71–78.

Holmes, S. J. "Are Recessive Characters due to Loss?" *Science*, n.s. 42 (1915): 300–3.

——. *Studies in Evolution and Eugenics.* New York: Harcourt Brace, 1923.

Hooker, Sir Joseph Dalton. "On the Origination and Distribution of Vegetable Species—Introductory Essay to the Flora of Tasmania." *Am. J. Sci.*, 2d ser. 29 (1860): 1–25, 305–26.

Hutton, F. W. "The Place of Isolation in Organic Evolution.' *Natural Science* 11 (1897): 240–46.

Huxley, Julian S. "Constant Differential Growth-Ratios and Their Significance." *Nature* 114 (1924): 895–96.

——. *Problems of Relative Growth.* London: Methuen, 1932.

——. *Evolution: The Modern Synthesis.* London: Allen & Unwin, 1942. New ed., New York: Wiley, 1964.

Huxley, Leonard. *The Life and Letters of Thomas Henry Huxley.* 3 vols. 2d ed. London: Macmillan, 1908.

——. *The Life and Letters of Sir Jospeh Dalton Hooker, O.M., G.C.S.I.: Based on Materials Collected and Arranged by Lady Hooker.* 2 vols. London: Murray, 1918.

Huxley, Thomas Henry. *American Addresses: With a Lecture on the Study of Biology.* New York: Appleton, 1888.

——. *Collected Essays.* 9 vols. Vol. 2, *Darwiniana*, 1893. Vol. 8, *Discourses Biological and Geological,* 1894. Vol. 9, *Evolution and Ethics,* 1894. London: Macmillan, 1893–94.

——. *The Scientific Memoirs of Thomas Henry Huxley,* ed. Sir M. Foster and E. Ray Lankester. 4 vols. London: Macmillan, 1899.

Hyatt, Alpheus. "On the Parallelism between the Different Stages of Life in the Individual and those in the Entire Group of the Molluscous Order Tetrabranchiata." *Mem. Boston Soc. Nat. Hist.* 1 (1866): 193–209.

——. "The Genesis of the Tertiary Species of Planorbis at Steinheim." *Anniversary Memoir of the Boston Society of Natural History (1830-1880),* pp. 1–114. Summarized *Proc. Am. Assoc. Adv. Sci.* 29 (1880): 527–50, and *Am. Naturalist* 16 (1882): 441–55.

——. "Evolution of the Cephalopoda." *Science* 3 (1884): 122–27.

——. *Genesis of the Arietidae.* Smithsonian Contributions to Knowledge, no. 673. Washington, D. C., 1889.

——. "Bioplastology and Related Branches of Biologic Research." *Proc. Boston Soc. Nat. Hist.* 26 (1892): 59–125.

——. "Phylogeny of an Acquired Characteristic." *Proc. Am. Phil. Soc.* 32 (1893): 349–647.

——. "Lost Characteristics." *Am. Naturalist* 30 (1896): 9–17.

Hyatt, Alpheus. "Cycle of Life in the Individual (Ontogeny) and in the Evolution of the Group (Phylogeny)." *Proc. Am. Acad. Arts & Sci.* 32 (1897a): 209–24.

———. "The Influence of Woman in the Evolution of the Human Race." *Natural Science* 11 (1897b): 89–93.

Jenkin, Fleeming. "The Origin of Species." *North British Review* 46 (1867): 277–318.

Jennings. Herbert Spencer. *Behavior of the Lower Organisms.* New York: Columbia University Press, 1906.

Jollos, V. "Studien zum Evolutionsproblem. 1.Über die experimentelle Hervorragung und Steigerung von Mutationen bei *Drosophila melanogaster.*" *Biol. Central-blatt* 50 (1930): 541–54.

Jones, Henry Festing. *Charles Darwin and Samuel Butler: A Step Toward Reconciliation.* London: Fifield, 1911.

———. *Samuel Butler, Author of Erewhon (1835–1902): A Memoir.* 2 vols. London: Macmillan, 1919.

Jordan, David Starr. "The Origin of Species Through Isolation." *Science,* n.s. 22 (1905): 545–62.

———, and Kellogg, Vernon. *The Scientific Aspects of Luther Burbank's Work.* San Francisco: A. M. Robertson, 1909.

Jordan, Karl. "Reproductive Divergence not a Factor in the Evolution of New Species." *Natural Science* 12 (1898): 45–47.

———. "Der Gegensatz zwischen geographischer und nicht-geographischer Variationen." *Zeitschrift für wissenschaftliche Zoologie* 83 (1905): 151–210.

Kammerer, Paul. "Vererbung erzwungener Fortpflanzungsanpassungen. 3. Mitteilung. Die Nachkommen der nicht Brutpflegenden *Alytes obstetricans.*" *Archiv für Entwickelungsmechanik* 28 (1909): 447–546.

———. "Adaptation and Inheritance in the Light of Modern Experimental Investigation." In *Smithsonian Inst. Ann. Rpt.,* 421–41. Washington, D. C., 1912.

———. "Experimente über Fortpflanzung ... IV Mitteilung. Das Farbkleid des Feuersalamanders, *Salamandra maculosa* Laurenti, in seiner Abhängigkeit von der Umwelt." *Archiv für Entwickelungsmechanik* 36 (1913): 4–193.

———. "Breeding Experiments on the Inheritance of Acquired Characters." *Nature* 111 (1923): 637-40.

———. *The Inheritance of Acquired Characteristics,* trans. A. Paul Maerker-Brandon. New York: Boni & Liveright, 1924.

Kellogg, Vernon L. "Is There Determinate Variation?" *Science,* n.s. 24 (1906): 621–28.

———. *Darwinism Today: A Discussion of Present-Day Scientific Criticism of the Darwinian Selection Theories, together with a Brief Account of the Principal Other Proposed Auxiliary and Alternative Theories of Species-Formation.* New York: Henry Holt; London: George Bell, 1908.

———, and Bell, R. G. "Variations Induced in Larval, Pupal and Imaginal Stages of *Bombyx mori* by Controlled Varying Food Supply." *Science,* n.s. 18 (1903): 741-48.

Kelvin, William Thomson, Baron. *Popular Lectures and Addresses.* Vol. 2, *Geology and General Physics.* London: Macmillan, 1894.

Kidd, Walter. *Use-Inheritance, Illustrated by the Direction of Hair on the Bodies of Animals.* London: A. & C. Black, 1901.

———. *Initiative in Evolution.* London: H. F. Witherby, 1920.

Koken, Ernst. Palaeontologie und Descendenzlehre. Jena: G. Fischer, 1902.

Kropotkin, Peter. "The Direct Action of the Environment on Plants." *Nineteenth Century and After* 68 (1910): 58–77.

——. "Inheritance of Acquired Characters." *Nineteenth Century and After*, 71 (1912): 511–31.

——. *Mutual Aid: A Factor in Evolution.* 1914 ed. Reprint, intro. Ashley Montagu. Boston: Extending Horizon Books, n.d.

——. "The Direct Action of Environment and Evolution." In *Smithsonian Inst. Ann. Rept.*, 409–27. Washington, D.C., 1918. Also *Nineteenth Century and After* 85 (1919): 70–89.

Lamarck, J. B. P. A. de Monet, Chevalier de. *Philosophie zoologique: Ou exposition des considerations relatives a l'histoire naturelle des animaux.* New ed., intro. Charles Martins. 2 vols. Paris: Savy, 1873.

——. *Zoological Philosophy: An Exposition with Regard to the Natural History of Animals*, trans. Hugh Elliot. London, 1914. Reprint. New York: Hafner, 1963.

Lang, W. D. "The Pelmatoprinae: An Essay on the Evolution of a Group of Cretaceous Polyzoa." *Phil. Trans. Roy. Soc. Lond.* (B), 209 (1919): 191–228.

——. *Catalogue of the Fossil Bryozoa (Polyzoa) in the Department of Geology, British Museum (Natural History). The Cretaceous Bryozoa*, vol. 3. London, 1921.

——. "Evolution: A Resultant." *Proc. Geol. Assoc.* 34 (1924): 7–20.

Lankester, E. Ray. "Inheritance of Acquired Characters." *Nature* 39 (1888–89): 485.

——. *The Advancement of Science: Occasional Essays and Addresses.* London: Macmillan, 1890.

——. "Acquired Characters." *Nature* 51 (1894–95a): 54, 102–3.

——. "Are Specific Characters Useful?" *Nature* 54 (1894–95b): 245–46.

——. "President's Address." *Report of the British Association for the Advancement of Science*, 1906 meeting, pp. 3–42.

Leconte, Joseph. *Evolution: Its Nature, its Evidences and its Relation to Religious Thought.* 2d ed. New York: Appleton, 1899.

Le Dantec, Félix. *Lamarckiens et Darwiniens: Discussion de quelques théories sur la formation des espèces.* Paris: Alcan, 1899.

——. *Eléments de la philosophie biologique.* Paris: Alcan, 1907a.

——. *The Nature and Origin of Life in the Light of New Knowledge*, trans. Stoddard Dewey. London: Hodder & Stoughton, 1907b.

Lock, R. H. *Recent Progress in the Study of Variety, Heredity and Evolution.* London: Murray, 1907.

Loomis, F. B. "Momentum in Variation." *Am. Naturalist* 39 (1905): 839–43.

Lotka, Alfred J. "Contributions to the Energetics of Evolution." *Proc. Nat. Acad. Sci.* 8 (1922); 147–51.

Lotsy, J. P. *Evolution by Means of Hybridization.* The Hague: Martinus Nijhoff, 1916.

Lull, Richard Swann. *Organic Evolution.* New York: Macmillan, 1917. Rev. ed., 1936.

——. "Dinosaurian Climatic Response." In *Organic Adaptation to the Environment*, ed. M. R. Thorpe, 225–279. New Haven: Yale University Press, 1924.

Lyell, Sir Charles. *Geological Evidences of the Antiquity of Man: With Remarks on Theories of the Origin of Species by Variation.* London: Murray, 1863.

——. *Principles of Geology: Or the Modern Changes of the Earth and its Inhabitants Considered as Illustrative of Geology.* 2 vols. 10th ed. London: Murray, 1867–68.

MacBride, E. W. *An Introduction to the Study of Heredity.* London: Williams & Norgate, 1924.

——. *Evolution.* London: Ernest Benn, 1927.

——. "The Inheritance of Acquired Characters." *Nature* 129 (1932): 900; and 130 (1932): 128.

MacDougal, D. T. "Discontinuous Variation and the Origin of Species." *Science,* n.s. 21 (1905): 540–43.

MacDougall, William. "An Experiment for the Testing of the Hypothesis of Lamarck." *British Journal of Psychology* 17 (1927): 267–304.

Marchant, James. *Alfred Russel Wallace: Letters and Reminiscences.* New York: Harper, 1916.

Marsh, Othniel C. "Preliminary Description of *Hesperornis regalis* . . . ." *Am. J. Sci.,* 3d ser. 3 (1872): 360–65.

——. "On a New Sub-Class of Fossil Birds (Odontornithes)." *Am. J. Sci.,* 3d ser. 5 (1873): 161–62.

——. "Notice of Some New Equine Mammals from the Tertiary Formation." *Am. J. Sci.,* 3d ser. 7 (1874): 247–58.

——. "Notice of Some New Tertiary Mammals." *Am. J. Sci.,* 3d ser. 12 (1876a): 401–4.

——. "Recent Discoveries of Extinct Mammals." *Am. J. Sci.,* 3d ser. 12 (1876b): 59–61.

——. *Introduction and Succession of Vertebrate Life in North America.* New York: Appleton, 1878.

——. *Odontornithes: A Monograph on the Extinct Toothed Birds of North America.* Report of the Geological Exploration of the Fortieth Parallel, vol. 7. Washington, D. C., 1880.

——. *Dinocerata: A Monograph on an Extinct Order of Gigantic Mammals.* Monographs of the U.S. Geological Survey, vol. 10. Washington, D. C., 1886.

Matthew, W. D. "The Phylogeny of the Felidae." *Bull. Am. Mus. Nat. Hist.* 28 (1910): 289–316.

——. "The Evolution of the Horse: A Record and its Interpretation." *Quart. Rev. Biol.* I (1926): 139–85.

Mayr, Ernst. *Systematics and the Origin of Species.* New York: Columbia University Press, 1942.

Merriam, C. Hart. "Is Mutation a Factor in the Evolution of the Higher Vertebrates?" *Science,* n.s. 23 (1906): 241–56.

Merz, John Theodor. *A History of European Thought in the Nineteenth Century.* 4 vols. 3d ed. Edinburgh: Blackwood, 1908.

Metcalf, Maynard M. "Determinate Mutation." *Science,* n.s. 21 (1905): 355–56.

——. "Adaptation Through Natural Selection and Orthogenesis." *Am. Naturalist* 47 (1913a): 65–71.

——. "Trends in Evolution: A Discussion of Data Bearing on Orthogenesis." *J. Anat. & Physiol.* 45 (1913b): 1–45.

Metchnikoff, Elie. "La lutte pour l'existence entre les diverses parties de l'organisme." *Revue scientifique* 50 (1892): pp. 321–26.

Miller, Hugh. *Footprints of the Creator: Or the Asterolepis of Stromness.* 3d ed. London: Johnstone & Hunter, 1850.

Mivart, St. George Jackson. *On the Genesis of Species.* London: Macmillan, 1871.

——. "On the Development of the Individual and of the Species." *Proc. Zool. Soc. Lond.* (1884): 462–74.

———. "On the Possible Dual Origin of the Mammalia." *Proc. Roy. Soc. Lond.* 43 (1887-88): 372-79.

———. "Are Specific Characters the Result of Natural Selection?" *Nature* 54 (1896): 246-47.

Montgomery, Thomas H. *The Analysis of Racial Descent in Animals.* New York: Henry Holt, 1906.

Morgan, Conway Lloyd. *Habit and Instinct.* London: Methuen, 1896.

———. *Emergent Evolution. The Gifford Lectures Delivered in the University of St. Andrews in the Year 1922.* 2d ed. London: Williams and Norgate, 1927.

Morgan, Thomas Hunt. *Evolution and Adaptation.* 1903. Reprint. New York: Macmillan, 1908.

———. "Chance or Purpose in the Origin and Evolution of Adaptations." *Science,* n.s. 31 (1910): 201-10.

———. *A Critique of the Theory of Evolution.* Princeton: Princeton University Press, 1916.

———. *The Physical Basis of Heredity.* Philadelphia: J. B. Lippincott, 1919.

———. *Evolution and Genetics.* Princeton: Princeton University Press, 1925.

———. *The Theory of the Gene.* Rev. ed. New Haven: Yale University Press, 1928.

———. *The Scientific Basis of Evolution.* London: Faber & Faber, 1932.

Morse, Edward S. "Memorial of Professor Alpheus Hyatt." *Proc. Boston Soc. Nat. Hist.* 30 (1902): 413-33.

Müller, Fritz. *Für Darwin.* Leipzig: Engelmann, 1864.

———. *Facts and Arguments for Darwin,* trans. W. S. Dallas. London: Murray, 1869.

Nägeli, Carl. *Mechanische-physiologische Theorie der Abstammungslehre.* Munich: R. Oldenbourg, 1884.

———. *A Mechanico-Physiological Theory of Organic Evolution,* trans. V. A. Clark and F. A. Waugh. Chicago: Open Court, 1898.

Naudin, Charles. "Les espèces affines et la théorie de l'évolution." *Bull Soc. Bot. de France* 21 (1874): 240-72.

Nopsca, Francis, Baron. "Notes on Stegocephalia and Amphibia." *Proc. Zool. Soc. Lond.* (1930): 979-95.

Nordenskiöld, Erik. *The History of Biology: A Survey,* trans. L. B. Eyre. 1929. Reprint. New York: Tudor Publishing, 1946.

Orr, Henry B. *A Theory of Development and Heredity.* New York: Macmillan, 1893.

Osborn, Henry Fairfield. "Are Acquired Characters Inherited?" *Am. Naturalist* 25 (1889a): 191-216.

———. "Paleontological Evidence for the Transmission of Acquired Characters." *Am. Naturalist* 23 (1889b): 561-66.

———. "The Hereditary Mechanism and the Search for the Unknown Factors in Evolution." *Biological Lectures, Woods Hole* 3 (1894): 79-100. Also *Am. Naturalist,* 29 (1895): 418-39.

———. "The Limits of Organic Selection." *Am. Naturalist,* 31 (1897): 944-51.

———. *The Evolution of Mammalian Molar Teeth to and from the Triangular Type,* ed. W. K. Gregory. New York: Macmillan, 1907.

———. "The Four Inseparable Factors of Evolution. . . ." *Science,* n.s. 27 (1908): 148-50.

———. *The Age of Mammals in Europe, Asia and North America.* New York: Macmillan, 1910.

———. "The Continuous Origin of Certain Unit Characters as Observed by a Paleontologist." *Am. Naturalist* 46 (1912): 185-206, 249-78.

Osborn, Henry Fairfield. *The Origin and Evolution of Life on the Theory of Action, Reaction and Interaction of Energy.* New York: Scribners, 1917.

———. "William Bateson on Darwinism." *Science* 55 (1922): 194.

———. *Man Rises to Parnassus: Critical Epochs in the Prehistory of Man.* Princeton: Princeton University Press, 1927.

———. *From the Greeks to Darwin. The Development of the Evolution Idea through Twenty-Four Centuries.* 2d ed. New York: Scribners, 1929a.

———. *The Titanotheres of Ancient Wyoming, Dakota and Nebraska.* 2 vols. U.S. Geological Survey Monograph no. 55. Washington, D. C., 1929b.

———. *Cope: Master Naturalist. The Life and Letters of Edward Drinker Cope, with a Bibliography of his Writings Classified by Subject.* Princeton: Princeton University Press, 1931.

———. "The Nine Principles of Evolution Revealed by Paleontology." *Am. Naturalist* 66 (1932): 52-60.

———. "Aristogenesis: The Creative Principle in the Origin of Species." *Am. Naturalist* 68 (1934): 193-235.

Owen, Richard. *On the Nature of Limbs: A Discourse Delivered on Friday, February 9, at an Evening Meeting of the Royal Institution of Great Britain.* London: J. Van Voorst, 1849.

———. "Darwin on the Origin of Species." *Edinburgh Review* 111 (1860): 487-532.

———. "On the *Archaeopteryx* of von Meyer." *Phil. Trans. Roy. Soc. Lond.* 153 (1864): 33-47.

———. *On the Anatomy of the Vertebrates.* 3 vols. London: Longmans, Green, 1866-68.

Owen, Rev. Richard. *The Life of Richard Owen.* 2 vols. London: Murray, 1894.

Packard, Alpheus S., Jr. "On the Embryology of *Limulus polyphemus.*" *Proc. Am. Assoc. Adv. Sci.* 19 (1870): 247-55.

———. "On the Development of *Limulus polyphemus.*" *Mem. Boston Soc. Nat. Hist.* 2, pt. 2 (1872): 155-201.

———. "The Law of Evolution." *The Independent* 33 (5 February 1880): 10.

———. "On the Inheritance of Acquired Characters in Animals with a Complete Metamorphosis." *Proc. Am. Acad. Arts & Sci.* 29 (1884): 331-70.

———. "The Cave Fauna of North America." *Mem. Nat. Acad. Sci.* 4 pt. 1 (1888): 1-156.

———. "Hints on the Evolution of the Bristles, Spines and Tubercules of Certain Caterpillars." *Proc. Boston Soc. Nat. Hist.* 24 (1890): 494-500.

———. "On the Origin of the Subterranean Fauna of North America." *Am. Naturalist* 28 (1894): 727-51.

———. *Lamarck: The Founder of Evolution. His Life and Work.* New York: Longmans, Green, 1901.

———, and Putnam, F. W. *The Mammoth Cave and its Inhabitants.* Salem, Mass.: Salem Press, 1872.

Pearson, Karl. "Socialism and Natural Selection." *Fortnightly Review,* n.s. 56 (1894): 1-21.

———. "Regression, Heredity and Panmixia." *Phil. Trans. Roy. Soc. Lond.* (A), 197 (1896): 253-318.

———. "Mathematical Contributions to the Theory of Evolution: On the Law of Ancestral Heredity." *Proc. Roy. Soc. Lond.* 62 (1898): 386-412.

———. *The Grammar of Science.* 2d ed. London: A. & C. Black, 1900.

———. "On the Fundamental Conceptions of Biology." *Biometrika* 1 (1901): 320-44.

——. "Walter Frank Raphael Weldon." *Biometrika* 4 (1906): 1-52.

——. *The Life, Letters and Labours of Francis Galton.* 3 vols. in 4. Cambridge: Cambridge University Press, 1914-30.

Perrier, Edmond. "Le transformisme et les sciences physiques." *Revue scientifique* 16 (1879): 890-95,

——. *La philosophie zoologique avant Darwin.* Paris: Alcan, 1884.

——. *Lamarck.* Paris: Payot, 1925.

——, and Gravier, Charles. "La tachygénèse, ou accéleration embryogénique, son importance dans les modifications des phénomènes embryogéniques, son rôle dans la transformation des organismes." *Ann. des sci. nat. (zool.)* 16 (1902): 133-374.

Plate, Ludwig. *Über Bedeutung und Tragweite des Darwin'schen Selectionsprincips.* Leipzig: Engelmann, 1900.

——. *Über Bedeutung des Darwin'schen Selectionsprinzips und Probleme der Artbildung.* Leipzig: Engelmann, 1903.

——. *Selektionsprinzip und Probleme der Artbildung. Ein Handbuch des Darwinismus.* Leipzig: Engelmann, 1913.

Plough, H. H., and Ives, P. T. "Heat Induced Mutations in *Drosophila.*" *Proc. Nat. Acad. Sci.* 20 (1934): 268-73.

Poulton, Edward Bagnall. *The Colors of Animals: Their Meaning and Use, Especially Considered in the Case of Insects.* New York: Appleton, 1890.

——. "Acquired Characters." *Nature* 51 (1894-95): 126-27.

——. *Charles Darwin and the Theory of Natural Selection.* New York: Appleton, 1896.

——. *Essays on Evolution, 1889-1907.* Oxford: Oxford University Press, 1908.

——. "A Hundred Years of Evolution." *Report of the British Association for the Advancement of Science,* 1931 meeting, pp. 71-95.

Punnett, Reginald C. *Mendelism.* 2d ed. Cambridge: Macmillan and Bowes, 1907.

——. *Mimicry in Butterflies.* Cambridge: Cambridge University Press, 1915.

Radl, Emmanuel. *The History of Biological Theories,* trans. E. J. Hatfield. Oxford: Oxford University Press; London: Humphrey Milford, 1930.

Reid, G. Archibald. "The Inheritance of 'Acquired' Characters." *Nature* 77 (1908): 293-94, 391.

Rensch, Bernhard. *Das Prinzip geographischer Rassenkreise und das Problem der Artbildung.* Berlin: Borntraeger, 1929.

Ribot, T. *L'hérédité: Étude psychologique sur ses phénomènes, ses lois, ses conséquences.* Paris: Ladrange, 1873.

Rignano, Eugenio. *Upon the Inheritance of Acquired Characters: A Hypothesis of Heredity, Development and Assimilation,* trans. Basil C. Harvey. Chicago: Open Court, 1911.

Robson, G. C. *The Species Problem: An Introduction to the Study of Evolutionary Divergence in Natural Populations.* Edinburgh: Oliver & Boyd, 1928.

——, and Richards, O. W. *The Variation of Animals in Nature.* London: Longmans, Green, 1936.

Romanes, George John. *Mental Evolution in Animals. With a Posthumous Essay on Instinct by Charles Darwin.* London: Kegan Paul, Trench, 1883.

——. "Physiological Selection: An Additional Suggestion on the Origin of Species." *J. Linn. Soc. (Zool.)* 19 (1886): 337-411. Abstracted in *Nature* 34 (1886): 314-16., 336-40, 362-65.

——. "Physiological Selection." *Nineteenth Century* 21 (1887): 59-80.

Romanes, George John. *An Examination of Weismannism.* 2d ed. Chicago: Open Court, 1899.

———. *Darwin and After Darwin: An Exposition of the Darwinian Theory and a Discussion of Post-Darwinian Questions.* Vol. 1, *The Darwinian Theory.* 2d. ed. London: Longmans, Green, 1905. Vol. 2, *Post-Darwinian Questions. Heredity and Utility.* New ed. London: Longmans, Green, 1900. Vol. 3, *Post-Darwinian Questions. Isolation and Physiological Selection.* Chicago: Open Court, 1897.

Rostand, Jean. *Les chromosomes, artisans de l'hérédité et du sexe.* Paris: Hachette, 1928.

Russell, E. S. *Form and Function: A Contribution to the History of Animal Morphology.* London: Murray, 1916.

Ryder, John A. "On Like Mechanical (Structural) Conditions Producing Like Morphological Effects." *Am. Naturalist.* 12 (1878a): 157–60.

———. "On the Mechanical Genesis of Tooth Forms." *Proc. Acad. Nat. Sci. Philadelphia.* 30 (1878b): 45–80.

———. "On the Origin of Bilateral Symmetry and the Numerous Segments of the Soft Rays of Fishes." *Am. Naturalist* 13 (1879): 41–43.

———. "Proofs of the Effects of Habitual Use in the Modification of Animal Organisms." *Proc. Am. Phil. Soc.* 26 (1889): 541–49.

———. "A Physiological Hypothesis of Heredity and Variation." *Am. Naturalist* 24 (1890): 85–92.

———. "Dynamics in Evolution." *Biological Lectures, Woods Hole* 2 (1893a): 63–81.

———. "Energy as a Factor in Organic Evolution." *Proc. Am. Phil. Soc.* 31 (1893b): 192–203.

———. "The Inheritance of Modifications due to Disturbances of the Early Stages of Development, Especially in the Japanese Domesticated Races of Gold Carp." *Proc. Acad. Nat. Sci. Philadelphia* 45 (1893): 75–94.

———. "A Dynamical Hypothesis of Inheritance." *Biological Lectures, Woods Hole* 3 (1894): 25–54.

Salisbury, Robert Cecil, 3d Marquis. "Presidential Address." *Report of the British Association for the Advancement of Science,* 1894 meeting, pp. 3–15.

Schindewolf, O. H. *Paläontologie, Entwickelungslehre und Genetik: Kritik und Synthese.* Berlin: Bornträger, 1936.

Schuchert, Charles, and Levene, Clara Mae. *O. C. Marsh: Pioneer in Paleontology.* New Haven: Yale University Press, 1940.

Scott, William Berryman. "On the Osteology of *Mesohippus* and *Leptomeryx,* with Observations on the Modes and Factors of Evolution in the Mammalia." *Journal of Morphology* 5 (1891): 301–406.

———. "On Variations and Mutations." *Am. J. Sci.* 48 (1894): 355–74.

———. *The Theory of Evolution: With Special Reference to the Evidence upon which it is Founded.* New York: Macmillan, 1917.

———. *The History of Land Mammals in the Western Hemisphere.* New York: Macmillan, 1929.

———. *Memoirs of a Paleontologist.* Princeton: Princeton University Press, 1939.

Seeley, Harry Govier. "On the Classification of the Fossil Animals Commonly Named Dinosauria." *Proc. Roy. Soc. Lond.* 42 (1887–88): 165–71.

———. "On *Pareiasaurus bombidens* (Owen) and the Significance of its Affinities to Amphibians, Reptiles and Mammals." *Phil. Trans. Roy. Soc. Lond.* (B) 179 (1888a): 59–109.

———. "On Parts of the Skeleton of a Mammal from the Triassic Rocks of Klipfontein, Fraserberg, S. Africa (*Theriodesmus phylarchus*, Seeley), Illustrating the Reptilian Inheritance of the Mammalian Hand." *Phil. Trans. Roy. Soc. Lond.* (B), 179 (1888b): 141–55.

———. "The Reputed Mammals from the Karoo Formation of the Cape Colony." *Phil. Trans. Roy. Soc. Lond.* (B), 135 (1894): 1019–28.

Sellars, Roy Wood. *Evolutionary Naturalism.* 1929. Reprint. New York: Russell and Russell, 1969.

Semon, Richard. *Die Mneme als erhaltendes Prinzip im Wechsel des organischen Geschehens.* 3d ed. Leipzig: Engelmann, 1911.

———. *The Mneme.* London: Allen & Unwin; New York: Macmillan, 1921.

Semper, Karl. *The Natural Conditions of Existence as they Affect Animal Life.* London: Kegan Paul, 1881.

Shaler, Nathaniel Southgate. "Lateral Symmetry in Brachiopoda." *Proc. Boston Soc. Nat. Hist.* 15 (1872): 159–62.

———. *The Interpretation of Nature.* Boston: Houghton Mifflin, 1893.

———. *The Autobiography of Nathaniel Southgate Shaler: With a Supplementary Memoir by His Wife.* Boston: Houghton Mifflin, 1909.

Shaw, George Bernard. *The Bodley Head Bernard Shaw*, vol. 5. London: 1972.

Shull, A. Franklin. "Weismann and Haeckel: One Hundred Years." *Science* 81 (1934): 443–52.

———. *Evolution.* New York: McGraw-Hill, 1936.

Simpson, George Gaylord. *Tempo and Mode in Evolution.* New York: Columbia University Press, 1944.

———. "Biographical Memoir of William Berryman Scott." *Biog. Mem. Nat. Acad. Sci.* 25 (1948): 175–203.

Smuts, Jan Christiaan. *Holism and Evolution.* 1926. Reprint. New York: Viking, 1961.

Spencer, Herbert. *First Principles of a New Philosophy.* New York: Appleton, 1864a.

———. *Illustrations of Universal Progress: A Series of Discussions.* New York: Appleton, 1864b.

———. *Principles of Biology.* 2 vols. London: Williams & Norgate, 1864c.

———. *Essays: Scientific, Political and Speculative.* 3 vols. London: Williams & Norgate, 1883.

———. *The Factors of Organic Evolution.* London: Williams & Norgate, 1887.

———. "The Inadequacy of Natural Selection." *Contemporary Review* 43 (1893a): 153–66, 439–56.

———. "Professor Weismann's Theories." *Contemporary Review* 43 (1893b): 743–60.

———. *An Autobiography.* 2 vols. London: Williams & Norgate, 1904.

Spengler, Oswald. *The Decline of the West.* 1926–28. Reprint. New York: Knopf, 1939.

Standfuss, Max. "Synopsis of Experiments on Hybridization and Temperature made with Lepidoptera up to the end of 1898." *Entomologist* 30 (1900): 161–67, 283–92, 340–48; and 34 (1901): 11–13, 75–84.

Stirling, Keir B., ed. *Selected Works in Nineteenth-Century North American Paleontology.* New York: Arno, 1974.

Stockard, Charles R. "Experimental Modification of the Germ Plasm and its Bearing on the Inheritance of Acquired Characters." *Proc. Am. Phil. Soc.* 42 (1923): 311–25.

Stockard, Charles R., and Craig, Dorothy M. "An Experimental Study of the Influence of Alcohol on the Germ Cells and Developing Embryos of Mammals." *Am. Naturalist* 47 (1913): 641-82.

——, and Pananicolaon, George. "A Further Analysis of the Hereditary Transmission of Degeneracy and Deformation by the Descendants of Alcoholized Mammals." *Am. Naturalist* 50 (1916): 65-88, 144-77.

Sturtevant, A. H. "An Interpretation of Orthogenesis." *Science* 59 (1924): 579-80.

Swinnerton, Henry Hurd. *Outlines of Palaeontology.* London: Arnold, 1923.

Taylor, J. Lionel. "The Study of Variations." *Natural Science* 12 (1898): 231-38.

Thistleton-Dyer, W. T. "The Utility of Specific Characters." *Nature* 54 (1896): 293-94.

Thompson, D'Arcy Wentworth. *On Growth And Form.* Cambridge: Cambridge University Press, 1917. New ed., 1942.

Thomson, J. Arthur. *Heredity.* 2d ed. London: Murray, 1912.

——. *The Gospel of Evolution.* New York: Putnam, 1926.

Thorpe, Malcolm Rutherford, ed. *Organic Adaptation to the Environment.* New Haven: Yale University Press, 1924.

Vernon, H. M. "Reproductive Divergence: An Additional Factor in Evolution." *Natural Science* 11 (1897): 181-89.

——. *Variation in Animals and Plants.* London: Kegan Paul, Trench, Trubner, 1903.

Virchow, Rudolph. "Descendenz und Pathologie." *Virchows Archiv für Pathologisches Anatomie und Physiologie* 103 (1886): 1-14, 205-25, 413-36.

Waagen, Wilhelm. "Die Formenreihe des *Ammonites subradius.*" *E. W. Beneke's Geognostisch-Paläontologische Beiträge* 2 (1869): 179-256.

——. *Jurassic Fauna of Kutch.* Vol. 1, *Cephalopoda.* Palaeontologica Indica (Memoirs of the Geological Survey of India). Calcutta, 1875.

Wagner, Moritz. *Die Darwin'sche Theorie und das Migrationgesetz der Organismen.* Leipzig: Dunker & Humblot, 1868.

——. *The Darwinian Theory and the Law of the Migration of Organisms,* trans. J. L. Laird. London: E. Stanford, 1873.

——. *Die Entstehung der Arten durch räumliche Sonderung.* Basel: Schwabe, 1889.

Wallace, Alfred Russel. "On the Zoological Geography of the Malay Archipelago." *J. Linn. Soc.* 4 (1860): 172-84.

——. "On the Phenomena of Variation and Geographical Distribution as Illustrated by the Papilionidae of the Malayan Region." *Trans. Linn. Soc.* 25 (1864): 1-71.

——. *Contributions to the Theory of Natural Selection.* London: Macmillan, 1870.

——. "Natural Selection—Mr. Wallace's Reply to Mr. Bennett." *Nature* 3 (1870-71): 49-50.

——. *Darwinism: An Exposition of the Theory of Natural Selection, with Some of Its Applications.* London: Macmillan, 1889.

——. *The Malay Archipelago: The Land of the Orang-utan and the Bird of Paradise. A Narrative of Travel with Studies of Man and Nature.* Rev. ed., 1890. Reprint. New York: Dover, 1962.

——. "The Rev. George Henslow on Natural Selection." *Natural Science* 5 (1894): 177-83.

——. *Natural Selection and Tropical Nature.* New ed. London: Macmillan, 1895.

——. "The Problem of Utility: Are Specific Characters Always or Generally Useful?" *J. Linn. Soc. (Zool.)* 25 (1896): 481-96.

——. *Island Life: Or the Phenomena and Causes of Insular Faunas and Floras.* 3d ed. rev. London: Macmillan, 1911.

Watson, D. M. S. "Structure, Evolution and Origin of the Amphibia—The 'Orders' Rachitomi and Sterospondyli." *Phil. Trans. Roy. Soc. Lond.* (B), 209 (1919): 1-74.

——. "Croonian Lecture—The Evolution and Origin of the Amphibia." *Phil. Trans. Roy. Soc. Lond.* (B), 214 (1925-26): 189-256.

Weismann, August. *Über den Einfluss der Isolierung auf die Artbildung.* Leipzig: Englemann, 1872.

——. *Studien zur Descendenztheorie.* Leipzig: Engelmann, 1875.

——. *Studies in the Theory of Descent,* trans. R. Meldola. 2 vols. London: Sampson Low, 1880-82.

——. *Essays upon Heredity and Kindred Biological Problems,* ed. E. B. Poulton, Selmar Schönland, and Arthur E. Shipley. 2 vols. Oxford: Oxford University Press, 1891-92.

——. *Die Continuitat des Keimplasma als Grundlage einer Theorie der Vererbung: Ein Vortrag.* Jean: G. Fischer, 1892.

——. "The All-Sufficiency of Natural Selection." *Contemporary Rev.* 64 (1893a): 309-38, 596-610.

——. *The Germ Plasm: A Theory of Heredity,* trans. W. Newton Parker and Harriet Rönfeldt. London: Scott, 1893b.

——. *On Germinal Selection.* Chicago: Open Court, 1896. Also *The Monist* 6 (1895-96): 250-93.

——. *The Evolution Theory,* trans. J. Arthur Thomson and Margaret R. Thomson. 2 vols. London: Edward Arnold, 1902a.

——. *Vorträge über Descendenztheorie, gehalten an der Universität zu Freiburg im Breisgau.* 2d ed. Jean: Fischer, 1902b.

Weldon, W. F. R. "An Attempt to Measure the Death-rate due to the Selective Destruction of *Carcinas moenas* with Respect to Particular Dimensions." *Proc. Roy. Soc. Lond.* 57 (1894-95a): 360-79.

——. "Remarks on Variation in Animals and Plants." *Proc. Roy. Soc. Lond.* 57 (1894-95b): 379-82.

——. "The Utility of Specific Characters." *Nature* 54 (1896): 294-95.

——. "President's Address (Zoological Section)." *Report of the British Association for the Advancement of Science,* 1898 meeting, pp. 887-902.

——. "A First Study of Natural Selection in *Clausilia laminata.*" *Biometrika* 1 (1901a): 109-24.

——. "Mendel's Laws of Alternative Inheritance in Peas." *Biometrika* 1 (1901b): 228-53.

——. "Professor De Vries on the Origin of Species." *Biometrika* 1 (1901c): 365-74.

Whitehead, Alfred North. *Process and Reality: An Essay in Cosmology.* Cambridge: Cambridge University Press, 1929.

Whitman, Charles Otis. *Orthogenetic Evolution in Pigeons.* Vol. 1 of *The Posthumous Works of Charles Otis Whitman,* ed. Oscar Riddle. The Carnegie Institute Publication no. 257. Washington, D. C., 1919.

Wilberforce, Samuel. "Darwin's Origin of Species." *Quarterly Rev.* 108 (1860): 225-64.

Willey, A. *Amphioxus and the Ancestry of the Vertebrates.* Preface by H. F. Osborn. New York: Macmillan, 1894.

Willis, J. C. *Age and Area: A Study in Geographical Distribution and Origin of Species.* Cambridge: Cambridge University Press, 1922.

——. *The Course of Evolution.* Cambridge: Cambridge University Press, 1940.

Windle, Bertram. Review of Eimer, *Organic Evolution*. *Edinburgh Review* 172 (1890): 316–49.

Woodward, Arthur Smith. "President's Address (Geological Section)." *Report of the British Association for the Advancement of Science*, 1909 meeting, pp. 462–71.

Yule, G. Udney. "Mendel's Laws and Their Probable Relations to Intra-racial Heredity." *New Phytologist* 1 (1902): 193–207, 222–36.

Zittel, Karl von. "Palaeontology and the Biogenetic Law." *Natural Science* 6 (1895): 305–12.

———. *History of Geology and Palaeontology*. 2 vols. London: Scott, 1901.

SECONDARY SOURCES

Allen, Garland E. "Thomas Hunt Morgan and the Problem of Natural Selection." *J. Hist. Biology* 1 (1968): 113–39.

———. "Hugo De Vries and the Reception of the Mutation Theory." *J. Hist. Biology* 2 (1969a): 55–87.

———. "T. H. Morgan and the Emergence of a New American Biology." *Quart. Rev. Biology* 44 (1969b): 168–88.

———. "Opposition to the Mendelian-Chromosome Theory: The Physiological and Developmental Genetics of Richard Goldschmidt." *J. Hist. Biology* 7 (1974): 49–92.

———. *Life Sciences in the Twentieth Century*. New York: Wiley, 1975.

———. *Thomas Hunt Morgan: The Man and His Science*. Princeton: Princeton University Press, 1978.

Aronson, Lester R. "The Case of *The Case of the Midwife Toad*." *Behavior Genetics* 5 (1975): 115–25.

Aulie, Richard P. "The Origin of the Idea of the Mammal-like Reptile." *American Biology Teacher* 36 (1974): 476–84, 545–53; and 37 (1974): 21–32.

Bajema, Carl Jay. *Eugenics, Then and Now*. Stroudsberg, Pa.: Dowden, Hutchinson & Ross, 1977.

Bannister, Robert C. "'The Survival of the Fittest is our Doctrine'—History or Histrionics?" *J. Hist. Ideas* 31 (1970): 377–98.

———. *Social Darwinism: Science and Myth in Anglo-American Social Thought*. Philadelphia: Temple University Press, 1979.

Barnett, S. A., ed. *A Century of Darwin*. 1958. Reprint. London: Mercury Books, 1962.

Baroni-Urbani, Cesare. "Hologenesis, Phylogenetic Systematics, and Evolution." *Systematic Zoology* 26 (1977): 343–46.

Bartholomew, Michael. "Huxley's Defence of Darwin." *Annals of Science* 32 (1975): 525–35.

Beddall, Barbara G. *Wallace and Bates in the Tropics: An Introduction to the Theory of Natural Selection*. London: Macmillan, 1969.

Bell, P. R., ed. *Darwin's Biological Work*. 1959. Reprint. New York: Wiley, 1964.

Blum, Harold F. *Time's Arrow and Evolution*. 2d ed. Princeton: Princeton University Press, 1955.

Boesiger, Ernest. "Evolutionary Biology in France at the Time of the Evolutionary Synthesis." In *The Evolutionary Synthesis*, ed. Ernst Mayr and W. B. Provine, 309–22. Cambridge: Harvard University Press, 1980.

Bowler, Peter J. "Darwin's Concepts of Variation." *J. Hist. Medicine* 29 (1974): 196–212.

——. "The Changing Meaning of 'Evolution.'" *J. Hist. Ideas* 36 (1975): 95–114.

——. "Alfred Russel Wallace's Concepts of Variation." *J. Hist. Medicine* 31 (1976a): 17–29.

——. *Fossils and Progress: Paleontology and the Idea of Progressive Evolution in the Nineteenth Century.* New York: Science History Publications, 1976b.

——. "Darwinism and the Argument from Design: Suggestions for a Re-evaluation." *J. Hist. Biology* 10 (1977a): 29–43.

——. "Edward Drinker Cope and the Changing Structure of Evolution Theory." *Isis* 68 (1977b): 249–65.

——. "Hugo De Vries and Thomas Hunt Morgan: The Mutation Theory and the Spirit of Darwinism." *Annals of Science* 35 (1978): 55–73.

——. "Theodor Eimer and Orthogenesis: Evolution by Definitely Directed Variation." *J. Hist. Medicine* 34 (1979): 40–73.

Box, Joan Fisher. *R. A. Fisher: The Life of a Scientist.* New York: Waley, 1978.

Burchfield, Joe D. "Darwin and the Dilemma of Geological Time." *Isis* 65 (1974): 301–21.

——. *Lord Kelvin and the Age of the Earth.* New York: Science History Publications, 1975.

Burkhardt, Richard W., Jr. *The Spirit of System: Lamarck and Evolutionary Biology.* Cambridge: Harvard University Press, 1977.

——. "Lamarckism in Britain and the United States." In *The Evolutionary Synthesis,* ed. Ernst Mayr and W. B. Provine, 343–52. Cambridge: Harvard University Press, 1980.

Burrow, J. W. *Evolution and Society: A Study in Victorian Social Thought.* Cambridge: Cambridge University Press, 1966.

Bury, J. B. *The Idea of Progress: An Inquiry into its Growth and Origin.* 1932. Reprint. New York: Dover, 1955.

Cannon, H. G. *Lamarck and Modern Genetics.* Springfield, Ill.: Charles C. Thomas, 1959.

Carter, G. S. *A Hundred Years of Evolution.* London: Sidgwick & Jackson, 1958.

Churchill, Frederick B. "August Weismann and a Break from Tradition." *J. Hist. Biology* 1 (1968): 91–112.

——. "Hertwig, Weismann, and the Meaning of Reduction Division." *Isis* 61 (1970): 429–57.

——. "William Johannsen and the Genotype Concept." *J. Hist. Biology* 7 (1974): 5–30.

——. "Rudolph Virchow and the Pathologists' Criteria for the Inheritance of Acquired Characters." *J. Hist. Medicine* 31 (1976): 117–48.

——. "The Weismann-Spencer Controversy over the Inheritance of Acquired Characters." *Proceedings of the 15th International Congress of the History of Science* (1978): 451–68.

——. "The Modern Synthesis and the Biogenetic Law." In *The Evolutionary Synthesis,* ed. Ernst Mayr and W. B. Provine, 112–22. Cambridge: Harvard University Press, 1980.

Clarke, Ronald W. *JBS: The Life and Work of J. B. S. Haldane.* London: Hodder & Stoughton, 1968.

Cock, A. G. "William Bateson, Mendelism, and Biometry." *J. Hist. Biology* 6 (1973): 1–36.

Coleman, William. "Science and Symbol in the Turner Frontier Hypothesis." *Am. Hist. Rev.* 72 (1966): 22-49.

——. "Bateson and Chromosomes: Conservative Thought in Science." *Centaurus* 15 (1970): 228-314.

——. *Biology in the Nineteenth Century: Problems of Form, Function, and Transformation.* New York: Wiley, 1971.

Conry, Yvette. *Correspondence entre Charles Darwin et Gaston de Saporta. Precédé d'une histoire de la paléobotanique en France.* Paris: P. U. F., 1972.

——. *L'introduction du Darwinisme en France au XIXᵉ siècle.* Paris: Vrin, 1974.

Cowan, Ruth Schwartz. "Francis Galton's Contributions to Genetics." *J. Hist. Biology* 5 (1972): 389-412.

Cravens, Hamilton. *The Triumph of Evolution: American Scientists and the Heredity-Environment Controversy, 1900-1941.* Philadelphia: University of Pennsylvania Press, 1978.

Daniels, George. *Darwinism Comes to America.* Waltham, Mass.: Blaisdell, 1968.

——. *Science in American Society: A Social History.* New York: Knopf, 1971.

Darden, Lindley. "Reasoning in Scientific Change: Charles Darwin, Hugo De Vries, and the Discovery of Segregation." *Stud. Hist. Phil. Sci.* 7 (1976): 127-69.

——. "William Bateson and the Promise of Mendelism." *J. Hist. Biology* 10 (1977): 87-106.

De Beer, Sir Gavin. *Charles Darwin.* London: Nelson, 1963.

——. "Mendel, Darwin and Fisher." *Notes Roy. Soc. Lond.* 19 (1964): 192-226.

De Marrais, Robert, "The Double-edged Effect of Sir Francis Galton: A Search for the Motives of the Biometrician-Mendelian Debate." *J. Hist. Biology* 7 (1974): 141-74.

Desmond, Adrian J. *The Hot-blooded Dinosaurs: A Revolution in Paleontology.* New York: Dial Press, 1976.

Dexter, Ralph W. "The Development of A. S. Packard, Jr., as a Naturalist and Entomologist." *Bulletin of the Brooklyn Entomological Society* 52 (1957): 57-66, 101-12.

——. "The 'Salem Secession' of Agassiz Zoologists." *Essex Institute Historical Collections* 101 (1965): 27-39.

——. "The Impact of Evolutionary Theories on the Salem Group of Agassiz Zoologists." *Essex Institute Historical Collections* 115 (1979): 144-71.

Dronamraju, K. R., ed. *Haldane and Modern Biology.* Baltimore: Johns Hopkins Press, 1968.

Dupree, A. Hunter. *Asa Gray.* Cambridge: Harvard University Press, 1959.

Eiseley, Loren. *Darwin's Century: Evolution and the Men who Discovered It.* New York: Doubleday, 1958.

Ellegård, Alvar. *Darwin and the General Reader. The Reception of Darwin's Theory of Evolution in the British Periodical Press, 1859-72.* Göteburg: Acta Universitatis Gothenburgensis, 1958.

Farley, John. "The Initial Reaction of French Biologists to Darwin's *Origin of Species.*" *J. Hist. Biology* 7 (1974): 275-300.

Fichman, Martin, "Wallace: Zoogeography and the Problem of Land Bridges." *J. Hist. Biology* 10 (1977): 45-64.

Fothergill, Phillip G. *Historical Aspects of Organic Evolution.* London: Hollis and Carter, 1952.

Froggatt, P., and Nevin, N. C. "Galton's 'Law of Ancestral Heredity': Its Influence on the Early Development of Human Genetics." *History of Science* 10 (1971a): 1-27.

Froggatt, P., and Nevin, N. C. "The 'Law of Ancestral Heredity' and the Mendelian-Ancestrian Controversy in England." *J. Med. Genetics* 8 (1971b): 1-36.

Gasman, Daniel. *The Scientific Origins of National Socialism: Social Darwinism in Ernst Haeckel and the German Monist League.* New York: Elsevier, 1971.

Geison, Gerald L. "Darwin and Heredity: The Evolution of his Hypothesis of Pangenesis." *J. Hist. Medicine* 24 (1966): 375-411.

George, Wilma. *Biologist Philosopher: A Study of the Life and Writings of Alfred Russel Wallace.* New York: Abelard-Schuman, 1964.

Ghiselin, Michael T. *The Triumph of the Darwinian Method.* Berkeley and Los Angeles: University of California Press, 1969.

Gillespie, Neal C. "The Duke of Argyll, Evolutionary Anthropology, and the Art of Scientific Controversy." *Isis* 68 (1977): 40-54.

——. *Charles Darwin and the Problem of Creation.* Chicago: University of Chicago Press, 1979.

Gillispie, Charles C. *Genesis and Geology: A Study in the Relations of Scientific Thought, Natural Theology, and Social Opinion in Great Britain, 1790-1850.* 1951. Reprint. New York: Harper & Row, 1959.

——. *The Edge of Objectivity: An Essay in the History of Scientific Ideas.* Princeton: Princeton University Press, 1960.

Glick, Thomas F., ed. *The Comparative Reception of Darwinism.* Austin: University of Texas Press, 1972.

Gould, Stephen Jay. "Dollo on Dollo's Law: Irreversibility and the Status of Evolutionary Laws." *J. Hist. Biology* 3 (1970): 189-212.

——. "D'Arcy Thompson and the Science of Form." *New Literary History* 2 (1971): 229-58.

——. "On Biological and Social Determinism." *History of Science* 12 (1974a): 212-20.

——. "The Origin and Function of 'Bizarre' Structures: Antler Size and Skull Size in the 'Irish Elk,' *Megaloceros giganteus.*" *Evolution* 28 (1974b): 191-220.

——. "The Eternal Metaphors of Paleontology." In *Patterns of Evolution,* ed. A. Hallam, pp. 1-26. Amsterdam: Elsevier, 1977a.

——. *Ontogeny and Phylogeny.* Cambridge: Harvard University Press, 1977b.

——. "G. G. Simpson, Paleontology, and the Modern Synthesis." In *The Evolutionary Synthesis,* ed. Ernst Mayr and W. B. Provine, 153-72. Cambridge: Harvard University Press, 1980a.

——. "Is a New and General Theory of Evolution Emerging?" *Paleobiology* 6 (1980b): 119-30.

——. "Punctuated Equilibrium—A Different Way of Seeing." *New Scientist* 94 (1982): 137-41.

——, and Eldredge, Niles. "Punctuated Equilibria: The Tempo and Mode of Evolution Reconsidered." *Paleobiology* 3 (1977): 115-51.

——, and Lewontin, R. C. "The Spandrels of San Marco and the Panglossian Paradigm: A Critique of the Adaptationist Programme." *Proc. Roy. Soc. Lond.* (B), 205 (1979): 581-98.

Greene, John C. *The Death of Adam: Evolution and its Impact on Western Thought.* Ames: Iowa State University Press, 1959.

——. "The Kuhnian Paradigm and the Darwinian Revolution in Natural History." In *Perspectives in the History of Science and Technology,* ed. Duane H. D. Roller, 3-25. Norman: University of Oklahoma Press, 1971.

——. "Darwin as a Social Evolutionist." *J. Hist. Biology* 10 (1977): 1-27.

Gregory, Frederick. *Scientific Materialism in Nineteenth Century Germany.* Dordrecht: D. Reidel, 1977.

Gruber, Howard E., and Barrett, Paul. *Darwin on Man: A Psychological Study of Scientific Creativity.* New York: E. P. Dutton, 1974.

Gruber, Jacob W. *A Conscience in Conflict: The Life of St. George Jackson Mivart.* New York: Columbia University Press, 1960.

Haller, John S., Jr. *Outcasts from Evolution: Scientific Attitudes of Racial Inferiority, 1859-1900.* Urbana: University of Illinois Press, 1971.

Haller, Mark H. *Eugenics: Hereditarian Attitudes in American Thought.* New Brunswick, N. J.: Rutgers University Press, 1963.

Hamburger, Viktor. "Evolutionary Theory in Germany: A Comment." In *The Evolutionary Syntheses,* ed. Ernst Mayr and W. B. Provine, 303-9. Cambridge: Harvard University Press, 1980.

Hardy, Sir Alistair C. *The Living Stream: A Restatement of Evolution Theory and its Relation to the Spirit of Man.* London: Collins, 1965.

Himmelfarb, Gertrude. *Darwin and the Darwinian Revolution.* London: Chatto & Windus, 1959.

Hodge, M. J. S. "England." In *The Comparative Reception of Darwinism,* ed. Thomas F. Glick, 3-31. Austin: University of Texas Press, 1972a.

——. "The Universal Gestation of Nature: Chambers' *Vestiges* and *Explanations.*" *J. Hist. Biology* 5 (1972b): 127-52.

Hofstadter, Richard. *Social Darwinism in American Thought.* Rev. ed. Boston: Beacon Press, 1955.

Holt, Niles R. "Ernst Haeckel's Monistic Religion." *J. Hist. Ideas* 32 (1971): 265-80.

Hull, David L. "Darwinism and Historiography." In *The Comparative Reception of Darwinism,* ed. Thomas F. Glick, 388-402. Austin: University of Texas Press, 1972.

——. *Darwin and his Critics: The Reception of Darwin's Theory of Evolution by the Scientific Community.* Cambridge: Harvard University Press, 1973.

——. "Sociobiology: A Scientific Bandwagon or a Traveling Medicine Show?" In *Sociobiology and Human Nature,* ed. M. S. Gregory, Anita Silvers, and Diane Sutch, 136-63. San Francisco: Jossey-Bass, 1978.

——. "The Limits of Cladism." *Systematic Zoology* 28 (1979): 416-40.

——; Tessner, Peter D.; and Diamond, Arthur M. "Plank's Principle: Do Younger Scientists Accept New Scientific Ideas with Greater Alacrity than Older Scientists?" *Science* 202 (1978): 717-23.

Irvine, W. *Apes, Angels and Victorians: A Joint Biography of Darwin and Huxley.* 1955. Reprint. Cleveland: Meridian, 1959.

Jepsen, Glenn L. "Selection, 'Orthogenesis,' and the Fossil Record." *Proc. Am. Phil. Soc.* 93 (1949): 479-500.

Joravsky, D. *The Lysenko Affair.* Cambridge: Harvard University Press, 1970.

Koestler, Arthur. *The Ghost in the Machine.* New York: Macmillan, 1967.

——. *The Case of the Midwife Toad.* London: Hutchinson, 1971.

——. *Janus: A Summing Up.* London: Hutchinson, 1978.

——, and Smythies, J. R., eds. *Beyond Reductionism: New Perspectives in the Life Sciences.* 1969. Reprint. Boston: Beacon Press, 1971.

Kottler, Malcolm Jay. "Alfred Russel Wallace, the Origin of Man, and Spiritualism." *Isis* 65 (1974): 145-92.

——. "Hugo De Vries and the Rediscovery of Mendel's Laws." *Annals of Science* 36 (1976): 517-38.

——. "Charles Darwin's Biological Species Concept and Theory of Geographic Speciation: The Transmutation Notebooks." *Annals of Science* 35 (1978): 275-97.

——. "Darwin, Wallace and the Origin of Sexual Dimorphism." *Proc Am. Phil. Soc.* 124 (1980): 203-26.

Kuhn, Thomas S. *The Structure of Scientific Revolutions.* 2d ed. Chicago: University of Chicago Press, 1970.

Lanham, Url. *The Bone Hunters.* New York: Columbia University Press, 1975.

Leeds, Anthony. "Darwinian and 'Darwinian' Evolutionism in the Study of Society and Culture." In *The Comparative Reception of Darwinism,* ed. Thomas F. Glick, 437-77. Austin: University of Texas Press, 1972.

Lesch, John E. "The Role of Isolation in Evolution: George J. Romanes and John T. Gulick." *Isis* 66 (1975): 483-503.

Limoges, Camille. "Natural Selection, Phagocytosis, and Preadaptation: Lucien Cuénot,1886-1901." *J. Hist. Medicine* 31 (1976): 176-214.

——. "A Second Glance at Evolutionary Biology in France." In *The Evolutionary Synthesis,* ed. Ernst Mayr and W. B. Provine, 322-28. Cambridge: Harvard University Press, 1980.

Lomax, Elizabeth. "Infantile Syphilis as an Example of Nineteenth Century Belief in the Inheritance of Acquired Characteristics." *J. Hist. Medicine* 34 (1979): 23-39.

Lovejoy, Arthur O. "Recent Criticism of the Darwinian Theory of Recapitulation: Its Grounds and its Initiator." In *Forerunners of Darwin: 1745-1859,* ed. Bentley Glass, Owsei Temkin, and William L. Straus, Jr., 438-58. 1959. Reprint. Baltimore: Johns Hopkins Press, 1968.

Løvtrup, Søren, *The Phylogeny of the Vertebrata.* New York: Wiley, 1977.

Ludmerer, Kenneth M. *Genetics and American Society: A Historical Appraisal.* Baltimore: Johns Hopkins Press, 1972.

Lurie, Edward. *Louis Agassiz: A Life in Science.* Chicago: University of Chicago Press, 1960.

MacLeod, Roy M. "Evolutionism and Richard Owen." *Isis* 56 (1965): 259-80.

Mandelbaum, Maurice. *History, Man and Reason: A Study in Nineteenth Century Thought.* Baltimore, Johns Hopkins Press, 1971.

Manier, Edward. *The Young Darwin and his Cultural Circle: A Study of the Influences which Shaped the Language and Logic of the Theory of Natural Selection.* Dordrecht: D. Reidel, 1977.

Mayr, Ernst. *Evolution and the Diversity of Life: Selected Essays.* Cambridge: Harvard University Press, 1976.

——. "Some Thoughts on the History of the Evolutionary Synthesis." In Mayr and Provine, *The Evolutionary Synthesis,* ed. Ernst Mayr and W. B. Provine, 1-48. Cambridge: Harvard University Press, 1980.

—— and Provine, William B. *The Evolutionary Synthesis: Perspectives on the Unification of Biology.* Cambridge: Harvard University Press, 1980.

McKinney, H. Lewis. *Wallace and Natural Selection.* New Haven: Yale University Press, 1972.

McPherson, Thomas. *The Argument from Design.* London: Macmillan, 1979.

Medvedev, Zhores. *The Rise and Fall of T. D. Lysenko.* New York: Columbia University Press, 1969.

Millhauser, Milton. *Just Before Darwin: Robert Chambers and Vestiges.* Middletown, Conn.: Wesleyan University Press, 1959.

Monod, Jacques. *Chance and Necessity: An Essay on the Natural Philosophy of Modern Biology,* Trans. Austryn Wainhouse. London: Collins, 1972.

Montagu, Ashley. *Darwin: Competition and Cooperation.* New York: Henry Schumann, 1952.

Montgomery, William M. "Germany." In *The Comparative Reception of Darwinism,* ed. Thomas F. Glick, 81–116. Austin: University of Texas Press, 1972.

Moore, James R. *The Post-Darwinian Controversies: A Study of the Protestant Struggle to Come to Terms with Darwin in Britain and America, 1870-1900.* Cambridge: Cambridge University Press, 1979.

Norton, B. J. "The Biometric Defense of Darwinism." *J. Hist. Biology* 6 (1973): 283–316.

——. "Biology and Philosophy: the Methodological Foundations of Biometry." *J. Hist. Biology* 8 (1975a): 85–93.

——. "Metaphysics and Population Genetics: Karl Pearson and the Background to Fisher's Multi-factorial Theory of Inheritance." *Annals of Science* 32 (1975b): 537–53.

O'Brien, Charles F. "*Eozoön canadense:* The Dawn Animal of Canada." *Isis* 61 (1970): 200–23.

——. *Sir William Dawson: A Life in Science and Religion.* Memoirs of the American Philosophical Society 84, 1971.

Olby, R. C. *The Origins of Mendelism.* London: Constable, 1966.

Oldroyd, D. R. *Darwinian Impacts: An Introduction to the Darwinian Revolution.* Milton Keynes: Open University Press, 1980.

Olmsted, J. M. D. *Charles Édouard Brown-Séquard: A Nineteenth-Century Neurologist and Endocrinologist.* Baltimore: Johns Hopkins Press, 1946.

Oppenheimer, Jane. "An Embryological Enigma in the *Origin of Species.*" In *Forerunners of Darwin, 1745-1859,* ed. Bentley Glass, Owsei Temkin, and William L. Straus, Jr., 292–322. 1959. Reprint. Baltimore: Johns Hopkins Press, 1968.

Ospovat, Dov. "The Influence of Karl Ernst von Baer's Embryology, 1828-1859: A Reappraisal in the Light of Richard Owen's and William B. Carpenter's Paleontological Application of von Baer's Law." *J. Hist. Biology* 9 (1976): 1–28.

Pastore, Nicholas. *The Nature-Nurture Controversy,* foreword by Goodwin Watson. New York: King's Crown Press, 1949.

Paul, Harry W. *The Edge of Contingency: French Catholic Reaction to Scientific Change from Darwin to Duhem.* Gainesville: University Presses of Florida, 1979.

Peel, J. D. Y. *Herbert Spencer: The Evolution of a Sociologist.* London: Heinemann, 1971.

Persons, Stow, ed. *Evolutionary Thought in America.* New York: George Braziller, 1956.

Pfeifer, Edward J. "The Genesis of American Neo-Lamarckism." *Isis* 56 (1965): 156–67.

——. "United States." In *The Comparative Reception of Darwinism,* ed. Thomas F. Glick, 168–206. Austin: University of Texas Press, 1972.

Plate, Robert. *The Dinosaur Hunters: Othniel C. Marsh and Edward D. Cope.* New York: D. McKay, 1964.

Provine, William B. *The Origins of Theoretical Population Genetics.* Chicago: University of Chicago Press, 1971.

——. "The Role of Mathematical Population Genetics in the Evolutionary Synthesis of the 1930s and 1940s." *Stud. Hist. Biology* 2 (1978): 167–92.

Rachootin, Stan, and Thomson, Keith. "Epigenetics, Paleontology, and Evolution." In *Evolution Today,* ed. G. G. E. Scudder and J. L. Reveal, 181-93. Pittsburgh: Hunt Institute for Botanical Documentation, Carnegie-Mellon University, 1981.

Rainger, Ronald. "The Continuation of the Morphological Tradition in American Paleontology, 1880-1910." *J. Hist. Biology* 14 (1981): 129-58.

Randall, J. Herman, Jr. "The Changing Impact of Darwin on Philosophy." *J. Hist. Ideas* 22 (1961): 435-62.

Rensch, Bernhard. *Evolution Above the Species Level.* London: Methuen, 1959.

——. "Historical Development of the Present Synthetic Neo-Darwinism in Germany." In *The Evolutionary Synthesis,* ed. Ernst Mayr and W. B. Provine, 284-303. Cambridge: Harvard University Press, 1980.

Rogers, James Allen. "Russian Opposition to Darwinism in the Nineteenth Century." *Isis* 65 (1974): 487-505.

Rudwick, Martin J. S. *The Meaning of Fossils: Episodes in the History of Paleontology.* 2d ed. New York: Science History Publications, 1976.

Ruse, Michael. *The Darwinian Revolution: Science Red in Tooth and Claw.* Chicago: University of Chicago Press, 1979.

——. *Darwinism Defended: A Guide to the Evolution Controversies.* Reading, Mass.: Addison-Wesley, 1982.

Russell, Bertrand. *Portraits from Memory and Other Essays.* London: Allen & Unwin, 1956.

Russett, Cynthia Eagle. *Darwin in America: The Intellectual Response, 1965-1912.* San Francisco: W. H. Freeman, 1976.

Schneider, Herbert W. *A History of American Philosophy.* New York: Columbia University Press, 1946.

Schweber, Sylvan S. "The Origin of the *Origin* Revisited." *J. Hist. Biology* 10 (1977): 229-36.

Searle, G. R. *Eugenics and Politics in Britain, 1900-1914.* Leiden: Noordhoff International, 1976.

Shor, Elizabeth. *The Fossil Feud between E. D. Cope and O. C. Marsh.* New York: Exposition Press, 1974.

Simpson, George Gaylord. *The Meaning of Evolution.* New Haven: Yale University Press, 1949.

——. "The Baldwin Effect." *Evolution,* 7 (1953): 110-17.

——. *The Major Features of Evolution.* New York: Columbia University Press, 1953.

——. *This View of Life: The World of an Evolutionist.* New York: Harcourt, Brace & World, 1964.

——. "Mesozoic Mammals Revisited." In *Early Mammals,* ed. D. M. Kermack and K. A. Kermack. Supp. I to *Zool. J. Linn. Soc.* 50 (1971).

——. "The Concept of Progress in Organic Evolution." *Social Research* 41 (1973): 28-51.

Smith, Roger. "Alfred Russel Wallace: Philosophy and the Nature of Man." *Brit. J. Hist. Sci.* 6 (1972): 177-99.

——. "The Human Significance of Biology: Carpenter, Darwin and the *Vera Causa.*" In *Nature and the Victorian Imagination,* ed. U. C. Knoepflmacher and G. B. Tennyson, 216-30. Berkeley and Los Angeles: University of California Press, 1977.

Stebbins, Robert E. "France." In *The Comparative Reception of Darwinism,* ed. Thomas F. Glick, 117-67. Austin: University of Texas Press, 1972.

Steele, E. J. *Somatic Selection and Adaptive Evolution: On the Inheritance of Acquired Characters.* Toronto: Williams & Wallace, 1979.

Stephens, Lester G. "Joseph LeConte's Evolutionary Idealism: A Lamarckian View of Cultural History." *J. Hist. Ideas* 39 (1978): 465-80.

Stocking, George. "Lamarckianism in American Social Thought." *J. Hist. Ideas* 23 (1962): 239-56.

Sturtevant, A. H. *A History of Genetics.* New York: Harper & Row, 1965.

Swinburne, R. G. "Galton's Law—Formulation and Development." *Annals of Science* 21 (1965): 15-31.

Swinton, William E. "Harry Govier Seeley and the Karoo Reptiles." *Bull. Brit. Mus. (Nat. Hist.). Historical Series* 3 (1962): 1-39.

Teilhard De Chardin, Pierre. *The Phenomenon of Man.* London: Collins, 1959.

Thompson, Ruth D'Arcy. *D'Arcy Wentworth Thompson: The Scholar Naturalist, 1860-1948.* London: Oxford University Press, 1958.

Turner, Frank Miller. *Between Science and Religion: The Reaction to Scientific Naturalism in Late Victorian England.* New Haven: Yale University Press, 1974.

Vorzimmer, Peter J. "Charles Darwin and Blending Inheritance." *Isis* 54 (1963): 371-90.

——. *Charles Darwin: The Years of Controversy. The Origin of Species and its Critics, 1859-1882.* Philadelphia: Temple University Press, 1970.

Waddington, C. H. *The Evolution of an Evolutionist.* Edinburgh: Edinburgh University Press, 1975.

Werfel, Alma Mahler. *And the Bridge is Love.* New York: Harcourt Brace, 1958.

White, Morton. *Social Thought in America: The Revolt Against Formalism.* New ed. Boston: Beacon, 1957.

Wiener, Phillip P. *Evolution and the Founders of Pragmatism.* Cambridge: Harvard University Press, 1949.

Wilkie, J. S. "Galton's Contributions to the Theory of Evolution, with Special Reference to his Use of Models and Metaphors." *Annals of Science* 11 (1955): 194-205.

Willey, Basil. *Darwin and Butler: Two Versions of Evolution.* London: Chatto & Windus, 1940.

Williams-Ellis, Amabel. *Darwin's Moon: A Biography of Alfred Russel Wallace.* London: Blackie, 1966.

Woodcock, George. *Henry Walter Bates: Naturalist of the Amazons.* London: Faber & Faber, 1969.

Young, Robert M. "The Impact of Darwin on Conventional Thought." In *The Victorian Crisis of Faith,* ed. Anthony Symondson,, 13-16. London: SPCK, 1970.

——. "Darwin's Metaphor: Does Nature Select?" *The Monist* 55 (1971a): 442-503.

——. "Evolutionary Biology: Then and Now." *Science Studies* 1 (1971b): 177-206.

Zirkle, Conway. "The Early History of the Ideas of the Inheritance of Acquired Characters and Pangenesis." *Trans. Am. Phil. Soc.,* n.s. 35 (1946): 91-151.

AUTHOR INDEX

SUBJECT INDEX